629.13431
C989
2nd ed.

3-17-93

30.00

HAWKEYE INST. OF TECHNOLOGY

3 7944 1001 4303 7

UNDERSTANDING AIRCRAFT STRUCTURES

SECOND EDITION

JOHN CUTLER

WITHDRAWN

OXFORD

BLACKWELL SCIENTIFIC PUBLICATIONS

LONDON EDINBURGH BOSTON

MELBOURNE PARIS BERLIN VIENNA

30554

© John Cutler 1981, 1992

Blackwell Scientific Publications
Editorial Offices:
Osney Mead, Oxford OX2 0EL
25 John Street, London WC1N 2BL
23 Ainslie Place, Edinburgh EH3 6AJ
3 Cambridge Center, Cambridge,
  Massachusetts 02142, USA
54 University Street, Carlton
  Victoria 3053, Australia

Other Editorial Offices:
Librairie Arnette SA
2, rue Casimir-Delavigne
75006 Paris
France

Blackwell Wissenschafts-Verlag
Meinekestrasse 4
D-1000 Berlin 15
Germany

Blackwell MZV
Feldgasse 13
A-1238 Wien
Austria

All rights reserved. No part of this publication
may be reproduced, stored in a retrieval system,
or transmitted, in any form or by any means,
electronic, mechanical, photocopying, recording
or otherwise without the prior permission of the
publisher.

First published in Great Britain by
Granada Publishing Ltd 1981
Reissued in paperback by
Collins Professional and Technical Books
1986
Reprinted 1988 by
BSP Professional Books
Second edition 1992

Set by DP Photosetting, Aylesbury, Bucks
Printed and bound in Great Britain
at the University Press, Cambridge.

DISTRIBUTORS

Marston Book Services Ltd
PO Box 87
Oxford OX2 0DT
(*Orders*: Tel: 0865 791155
          Fax: 0865 791927
          Telex: 837515)

USA
Blackwell Scientific Publications, Inc.
3 Cambridge Center
Cambridge, MA 02142
(*Orders*: Tel: 800 759-6102
          617 225-0401)

Canada
Oxford University Press
70 Wynford Drive
Don Mills
Ontario M3C 1J9
(*Orders*: Tel: 416 441-2941)

Australia
Blackwell Scientific Publications
(Australia) Pty Ltd
54 University Street
Carlton, Victoria 3053
(*Orders*: Tel: 03 347-0300)

British Library
Cataloguing in Publication Data
Cutler, John
  Understanding aircraft structures. – 2nd ed
  I. Title
  629.13431

ISBN 0–632–03241–3

Library of Congress
Cataloging in Publication Data
Cutler, John.
  Understanding aircraft structures/John
  Cutler. — 2nd ed.
    p.   cm.
  Includes bibliographical references
    and index.
  ISBN 0–632–03241–3
    1. Airframes.   I. Title.
  TL671.6.C88   1992
  629. 134′31—dc20   91-38352

Contents

Preface

In the ten years since the first publication of this book, there have been changes to many of the specifications and publications which were referred to by specific identity numbers; some others have disappeared or been replaced. So the first objective of the new edition is to revise the identities of important reference documents etc.

The second objective has been to incorporate some of the improvements suggested by the many people who have written with kind comments. They are too many to mention individually but their help is gratefully acknowledged.

The text on airworthiness engineering and quality control, which is one of the main additions suggested by readers, will give an initial 'overview' to young engineers, perhaps particularly helpful to those specialising in other disciplines, but who need an insight into these functions. There is also a new appendix on common errors of detail design with some suggestions on how to avoid their worst effects.

One area which has not been changed is the use of imperial units (rather than S.I.) in the text. The aircraft industry, due to the strong American influence, has been slow to adopt metrication and it was felt that there was some merit in keeping students in particular, aware of the old order.

CHAPTER 1

Introduction

All professions have their own words and phrases amounting to a language which is unique to that industry. Unfortunately this makes a barrier for the many people who wish to take (or need to take) an interest in arts or sciences which are outside their normal experience. For those who want to satisfy an intelligent curiosity in a subject by reading or by discussion with an expert, a knowledge of how to phrase questions in a resonable way and to have some appreciation of the answers must start with an understanding of the terms the expert uses to communicate ideas and difficulties.

This book explains aircraft structures in such a way, giving the reader an essential understanding of the language and its uses, and also illustrating some of the problems. It will, therefore, be of interest to people who work in other sectors of the aviation industry, in fields such as purchasing or accounting, or even to those who actually fly the aircraft. All these people have to listen, at times, to the structures engineer and this book will help them to understand what he is talking about.

It is also written for students who are just beginning their studies in aircraft engineering. Experience shows that there is a wide gap between the workbench and the lecture hall. The endeavour in these pages to breathe life into the process of mathematical analysis will help to provide a very useful bridge.

For ease of reference, the chapters are graded progressively in their treatment of technical detail and the reader is thus able to re-examine and clarify information at will before proceeding to the next stage.

The objective and scope of this book should be borne in mind. It is to impart the *basics* of aircraft structure and for many readers that will be sufficient for their purposes. An appropriate bibliography is included, however, for those with professional ambitions.

Readers who are already draughtsmen or licensed engineers will finish the book with sufficient understanding to make a start on the road to an engineering qualification in the subject, while students who are aiming to become aircraft designers or stress analysts (generically referred to as 'stressmen' in the text) will encounter explanations of phenomena which they will not, perhaps, easily find in more formal textbooks. Hopefully, an understanding of these principles will aid their advance.

Finally, it is necessary to explain that, throughout the world, the monitoring of work on aircraft structures is subject to the supervision and approval of the Airworthiness organisations of national aviation authorities; the most influential of these being the Federal Aviation Administration (FAA) of the United States and the Civil Aviation Authority (CAA) of the United Kingdom. By and large, the greater proportion of countries in the world aviation community has adopted similar standards and, for this reason, FAA and CAA rulings and procedures form the reference frame of this work.

CHAPTER 2

History

2.1 Outline

2.1.1

In this chapter we will look at the general development of aircraft structures over the short period of their history. As with most subjects a knowledge of the steps which led to the present position is a great help in understanding current problems; later in the book there are more detailed comments concerning structures as they are now.

Flying machines obviously changed enormously over the seventy years from the Wright Brothers Flyer at Kittyhawk to Apollo on the moon, and a fighter ace of 1918 flew a very different aircraft from that flown by his successor today, so a review of the whole development of flying would be a large task.

However, there are many different branches of science and engineering which make up aeronautics as a whole and when these are looked at separately, the problem of dealing with them becomes more manageable.

The main divisions of aeronautical engineering are (a) the science which deals with the airflow round the aircraft, (b) the power plant engineering, (c) the avionics (that is the radios and navigation aids), (d) the airframe engineering where the airframe includes hydraulic and electrical systems, flying and engine controls, interior furnishings and cargo systems; and the section which concerns this book, which is the structure. All these divisions and subdivisions have developed at different rates. Power plants (engines) for instance have moved with two great strides, and many years of continuing short but rapid steps. Before the Wright Brothers could make their first successful aeroplane, the power plant engineers had to make their first stride and invent an engine which was light and powerful. The next stride was the invention of the jet engine, but in between, the power of piston engines increased nearly 200 times in just forty years from 12 hp (horse power) to over 2000 hp, with only a ten times increase of weight. As we shall see, structures have made only one major fundamental jump forward, but that was sufficient to change the whole character and appearance of aircraft.

2.2 Wire Braced Structures

2.2.1

If we look at the aircraft in fig. 2.1 we can have no doubt about the form of its construction. The wings and the fore and aft structures carrying the other covered

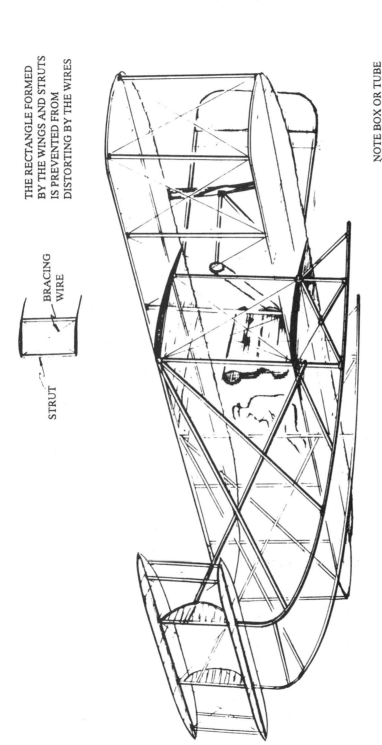

THE RECTANGLE FORMED
BY THE WINGS AND STRUTS
IS PREVENTED FROM
DISTORTING BY THE WIRES

BRACING
WIRE

STRUT

NOTE BOX OR TUBE
FORMED WITH BIPLANE
WINGS

Fig. 2.1   Wright Flyer – 1903

surfaces were all made of rectangular frames which were prevented from collapsing (or parallelogramming) by wires stretched from corner to corner. Although the methods were not original, there were two imaginative pieces of structural thinking here. Firstly, the idea that two wings, one above the other, would make a lighter, stronger structure than the type of wing arrangement suggested by bird flight, and secondly, the idea that a rectangle could be held in shape with two light wires rather than with one much heavier diagonal member like a farm gate. At this stage, and for the next thirty years, the major structural material was wood; at first bamboo and later mainly spruce, a lightweight timber with very straight grain and medium strength. Strangely enough, balsa wood, which means so much to the model aircraft enthusiast, was not used during this period but has been used sometimes since then as a filler, or core material, in flooring panels for large aircraft. Wire bracing continued to be used as a major feature of aircraft construction for many years. Figs. 2.2 and 2.3 illustrate its extensive use on early fighters and fig. 2.4 shows it still in evidence into the era of metal aircraft. (Note: these particular aircraft are illustrated as they show a progression of designs from the same manufacturer; Sopwith Aviation Co. becoming Hawker's in 1920.)

2.2.2

Biplane structures (figs. 2.1, 2.2 and 2.3) dominated aircraft design for many years and, for certain particular requirements such as aerobatic aircraft or agricultural crop sprayers, they still appear from time to time. The structural advantage of the arrangement is that the combination of upper and lower wing, the interplane vertical members (struts) and the wire bracing, form together a deep light member which is very rigid and resistant to bending and twisting.

2.2.3

The biplane era lasted until the middle of the 1930s by which time wooden construction was being replaced by metal. The similarity of construction between the biplanes in figs. 2.2 and 2.3 is obvious but the main fuselage members of the later aircraft are steel tubes as are the wing spars. Although wood was still being used very extensively at the time of the Hawker Fury, by the time that the Hurricane was produced (see fig. 2.4), the change to metal was almost complete. In spite of this change, the structures were still wire braced, and the principles of structural thinking behind the Sopwith Camel of 1917 and the Hurricane of 1935 showed some distinct similarities. Although the structures designer had been pushed into accepting the problems of designing monoplane wings as part of the progress towards better flying performance, the main fuselage structure of both aircraft was four longitudinal members called longerons and wire bracing was still being used extensively in the Hurricane. The side panels of the later fuselage were braced on the 'Warren girder' principle, but this could not be called a great leap forward. This particular aircraft was designed at a period when some very rapid advances were being made in aircraft performance, and expansion of the airforces of the world was clearly preparing for the coming Second World War, so some safe conservative thinking on the design of the structure was understandable. However, the difficulty of combining metal construction with traditional fabric covering in a monoplane wing produced the strange design shown in fig. 2.4(a). This was very short lived and all but the first few Hurricanes had metal covered wings, but it is worth seeing how the unfamiliar problems of preventing bending and twisting in a monoplane wing were tackled by one

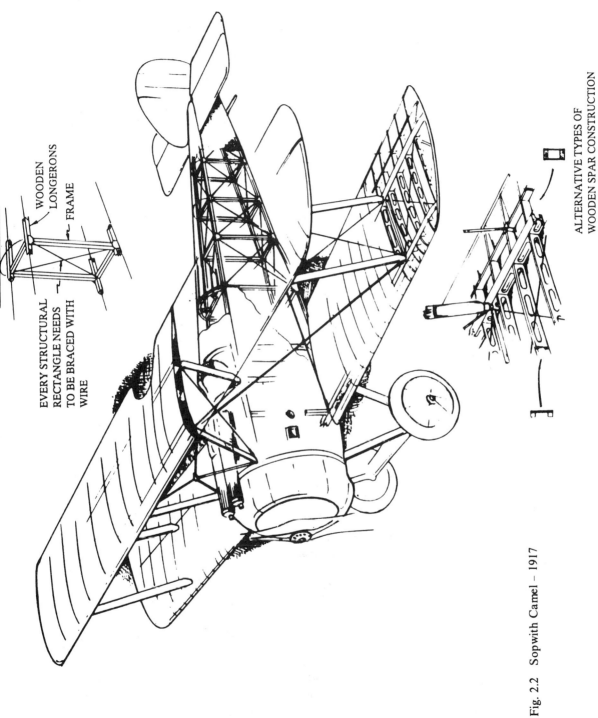

WOODEN LONGERONS

FRAME

EVERY STRUCTURAL RECTANGLE NEEDS TO BE BRACED WITH WIRE

ALTERNATIVE TYPES OF WOODEN SPAR CONSTRUCTION

Fig. 2.2   Sopwith Camel – 1917

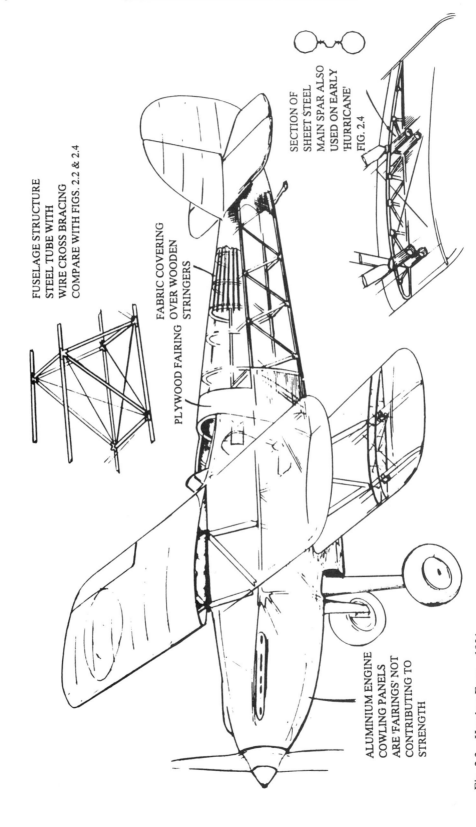

SECTION OF
SHEET STEEL
MAIN SPAR ALSO
USED ON EARLY
'HURRICANE'
FIG. 2.4

FUSELAGE STRUCTURE
STEEL TUBE WITH
WIRE CROSS BRACING
COMPARE WITH FIGS. 2.2 & 2.4

FABRIC COVERING
PLYWOOD FAIRING OVER WOODEN
STRINGERS

ALUMINIUM ENGINE
COWLING PANELS
ARE 'FAIRINGS' NOT
CONTRIBUTING TO
STRENGTH

Fig. 2.3   Hawker Fury – 1931

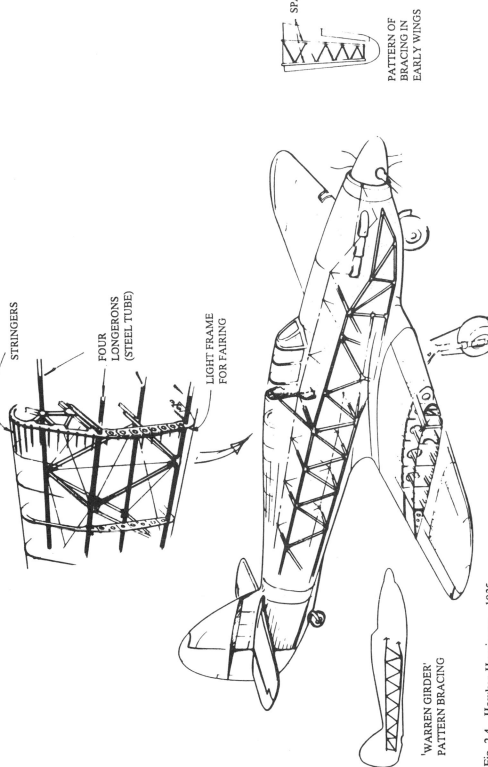

SPARS

PATTERN OF
BRACING IN
EARLY WINGS

STRINGERS

FOUR
LONGERONS
(STEEL TUBE)

LIGHT FRAME
FOR FAIRING

'WARREN GIRDER'
PATTERN BRACING

Fig. 2.4   Hawker Hurricane – 1935

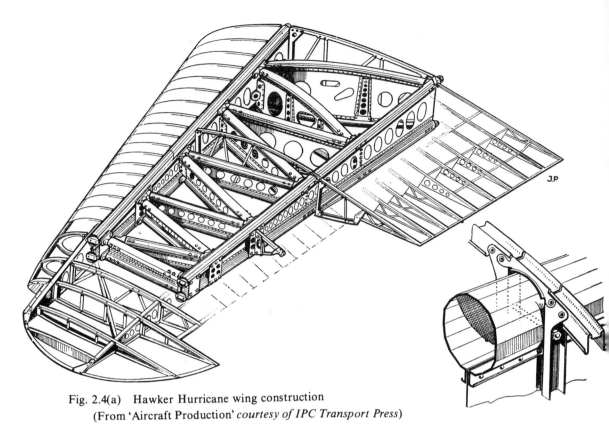

Fig. 2.4(a)   Hawker Hurricane wing construction
(From 'Aircraft Production' *courtesy of IPC Transport Press*)

designer in the interim period between biplane construction and the type of structure
we use today.

2.3   Semi Monocoque Structures

2.3.1

An essential feature of the fuselages discussed so far was the internal cross-bracing
which although acceptable for some types of aircraft was definitely in the way when
fuselages became large enough to carry passengers seated internally. A much more
suitable type of construction had in fact been flying for some while. When we look
back into aviation history, we refer to the flying boat without perhaps recognising just
how well it was named. The fuselages of the early wooden flying boats were made to
the highest standards of yacht construction with bent wood frames and double or
triple ply skins, carvel built (smooth exterior) with clear varnished finish. This method
of construction presented a much more open and usable fuselage interior. The wooden
frames and plywood skins passed from flying boats to land planes and the particular
building technique produced some quite shapely aeroplanes compared with the very
square 'slab sided' shapes which were the hall mark of the wire braced fuselage. The
main advantage of the boat type of construction from the engineering point of view
was, and is, that the skin forms an integral working part of the structure, unlike the

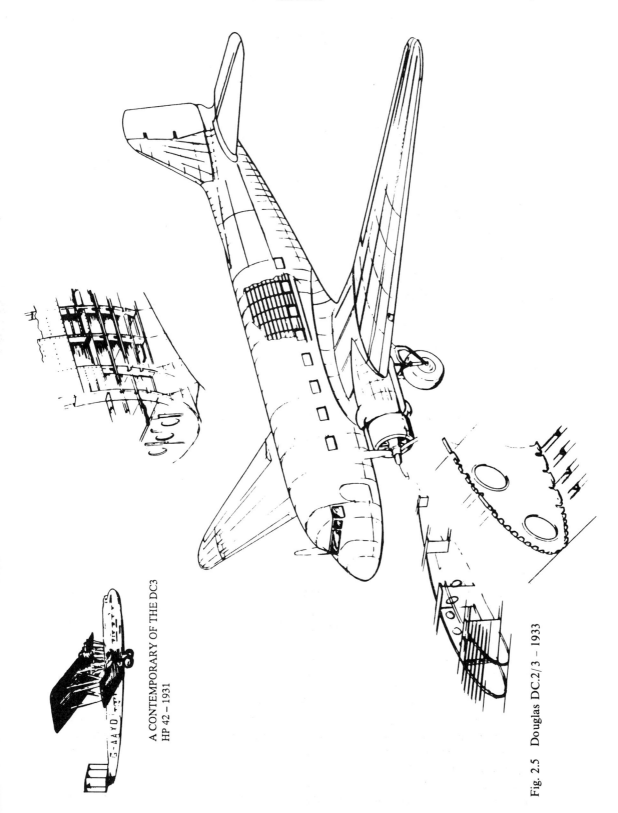

A CONTEMPORARY OF THE DC3
HP 42 – 1931

Fig. 2.5 Douglas DC.2/3 – 1933

covering of the braced fuselage which is a complete framework and which would still
be just as strong without the fabric. Because the skin is working, the boat construction
came to be called *stressed skin*. It is also sometimes called *semi monocoque*. (Coque is
the French word for eggshell and also the word meaning hull (of boats). Mono in
French implies 'all in one piece' or 'integral'. Semi is an English addition because the
construction uses an internal framework which is not a feature of pure monocoque.)

The adoption of this type of construction, with its material changed from wood to
metal, constituted the one major fundamental stride forward that aircraft structures
have taken since the earliest days of flying. The simplicity and elegance of this method
of design and the influence it brought to bear on the shape and appearance of aircraft is
shown in fig. 2.5. The Douglas DC.3 was not the first all-metal stressed skin aircraft,
but it was a brilliant example of a new technology converted into successful
engineering; it set standards and methods of structural design which have already
lasted for decades and look as if they will go on for generations.

2.3.2

After the fuselage shells came the stressed skin wings. The American aviation industry,
by 1930, was already showing signs of dominance in the building of civil aircraft and
was gaining stressed skin 'know how' with large capacity fuselages. European industry
was more concerned with military flying and secrecy, and continued to design braced
structures for perhaps two generations of aircraft (about five to seven years at that
time) longer than they should have done. To be fair, there was a problem of size. If a
DC.3 wing was scaled down to a size suitable for a fighter aircraft, the sheet aluminium
used for the construction tended to become unmanageably thin and flimsy. However,
some European design offices, notably Messerschmitt and Supermarine, faced up to
the problem and were making efficient stressed skin wings for small aircraft by the mid
1930s.

2.3.3

The later illustrations in this chapter, figs. 2.6 onwards, show the mainstream of
structural development since the DC.3. There has been a number of different
structural methods invented over the years but the arrangement of vertical members
(called ribs in the wing and frames or rings in the fuselage) and longitudinal members
(called stringers) has been the established convention for some years.

A glance at these post 1955 designs shows very clearly that the structures of big
aircraft are not just little aircraft structures scaled up (or the opposite way round). In
fact, whatever the size of the aircraft, the fuselage frames are always about 20 in. (500
mm) apart and have between 3 in. and 6 in. deep cross-sections. This means that big
aircraft have many structural members and small aircraft have few. This situation is
not altogether what one would expect, and although scaling up a small structure to
make a large one would give a strange result, scaling down large to small may be seen
to have more possibilities when the resultant small spaces between frames and
stringers are compared with the centre cores of the composite materials mentioned in
para 2.4.

Some other differences between the aircraft illustrated are discussed in chapter 9,
and we will restrict this chapter to saying that the developments which have taken
place in recent years have been mainly towards (a) a reduction of the number of rivets
in the aircraft, either by machining large pieces of structure from solid (see fig. 9.12), or

by the adhesive bonding assembly of components and (b) reducing the effects of minor structural damage, either by providing sufficient members so that the failure of one is not disasterous (fail safe structure), or by improving access for easy inspection of structures in service.

The methods of design and manufacture for this stressed skin type of construction are now so well understood by the manufacturers that the design of the structure for a new medium size, medium performance airliner could be regarded as a routine exercise, but the constant commercial pressure to reduce weight provides a strong incentive for improvement. Another major interest and area of innovation for the structures engineer is the search for methods of improving the reliability of structures, that is, reducing the time spent on the ground when minor structural defects are looked for and repaired. The British Aerospace 146 (fig. 2.10) is very interesting in this respect and the designers have taken a great deal of trouble to reduce the number of members and the number of places where structural damage can start. Some further comment will be found in chapter 9.

2.4   Sandwich Structures

2.4.1

One of the main problems and limitations of the skin material in stressed skin construction is its lack of rigidity. As we shall see later in this book, skins often have to be made thicker than they might otherwise need to be because of a tendency to crumple under some types of load. A strip of paper illustrates this problem very well, one can pull it but not push it.

A way of providing thin sheets with rigidity is to make a sandwich with one very thin sheet, a layer of very light but fairly rigid 'core' material, and another very thin sheet, all bonded together with an appropriate adhesive. As with conventional semi monocoque structures, wooden construction led the way with sandwich structures. The famous and elegant De Havilland Mosquito of 1940 was built with plywood skins either side of a balsa wood core. In today's major structures, a metal core of honeycomb-like cells is recognised as the most suitable core for metal faced sandwich (see figs. 2.13 and 2.14).

2.4.2

Although the advantages of honeycomb sandwich have been recognised for many years and flat boards are effectively a standard product, shaped products such as those in fig. 2.4(b) are difficult to make and therefore expensive. Nevertheless as fig. 2.14(c) shows, an appreciable percentage of the Boeing 747 structure is sandwich construction, and we can anticipate that the percentage will increase in later designs.

A British innovation some years ago avoided one of the problems of straightforward honeycomb core material. By a modification of the hexagon pattern a core was made which could be formed round panels which were double curved, i.e. dome shaped. This facility is not possible with conventional core which has to be carved to shape from a large block.

The rate of adoption of honeycomb sandwich is slower than many structures designers would wish. The concept of shell structures that are as smooth inside as they are outside and small fuselages with a ratio of structure thickness to outside diameter, as good as that of large fuselages, was a dream rather than a reality when this book was first published in 1981. Since then the Beech Starship, which is clearly highly

Fig. 2.6(a)   *(Courtesy of Beech Aircraft Corporation)*

Fig. 2.6(b)   *(Courtesy of Beech Aircraft Corporation)*

innovative in many ways, has been designed with a fuselage which exactly displays the advantages of a honeycomb shell.

Compare the photograph here with fig. 5.2 and the extent of the advance which has been made becomes apparent. Innovation on this scale clearly involves massive investment in research, design, tooling and courage and although the existing structural conventions will remain for some time, the Starship fuselage provides an elegant example of the benefits of sandwich construction.

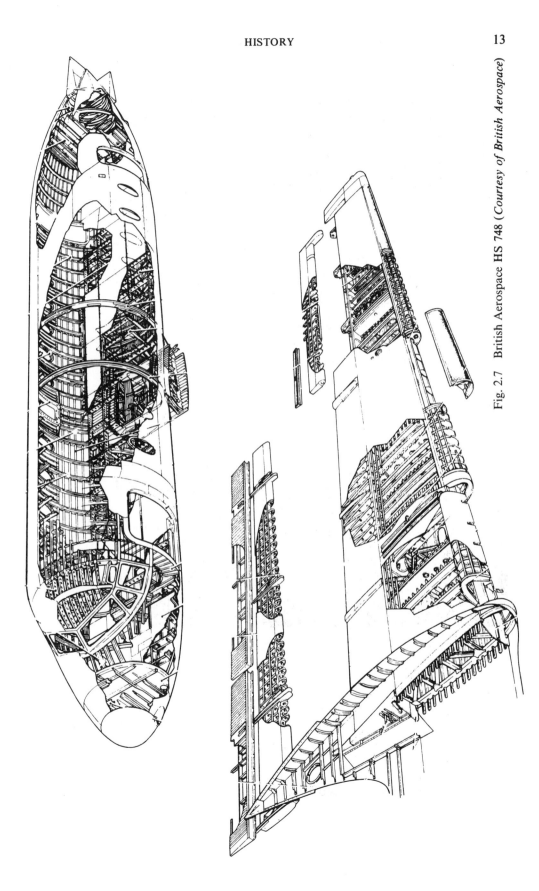

Fig. 2.7 British Aerospace HS 748 (*Courtesy of British Aerospace*)

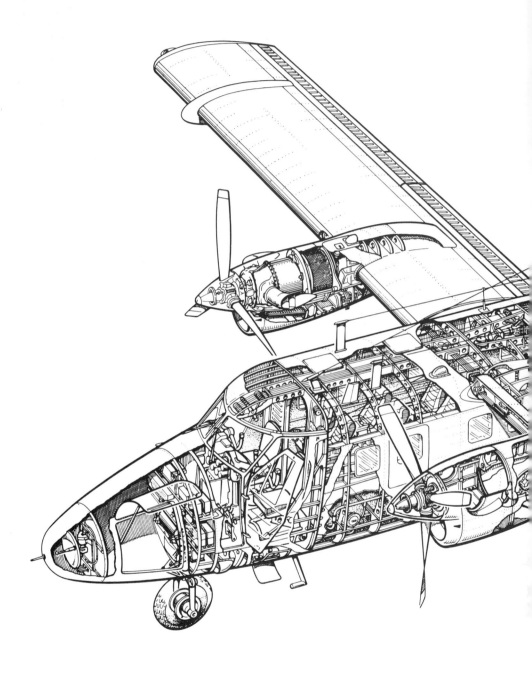

Fig. 2.8(a) De Havilland Canada – Twin Otter *(Courtesy of The De Havilland Aircraft of Canada Ltd)*

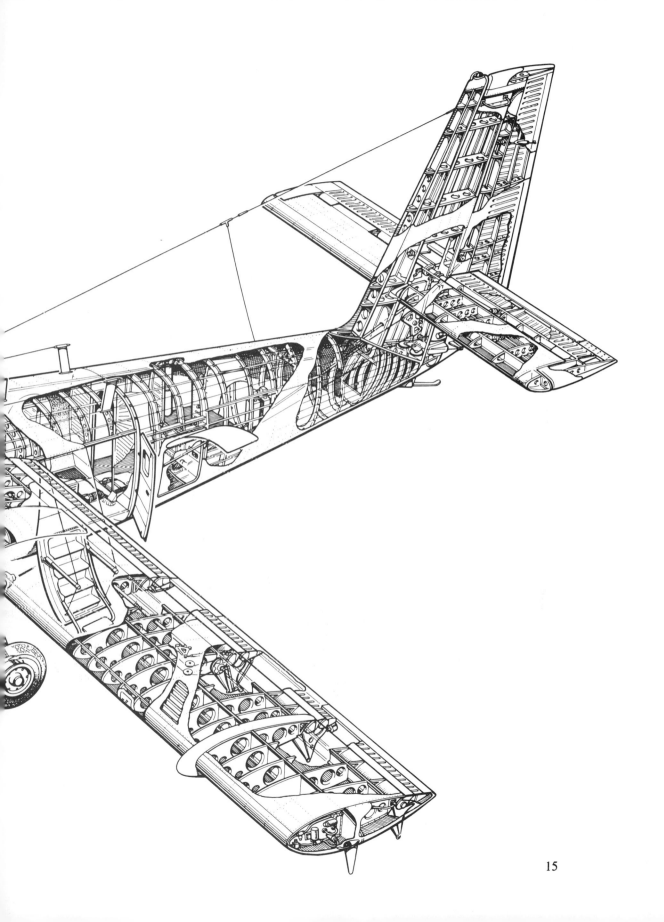

15

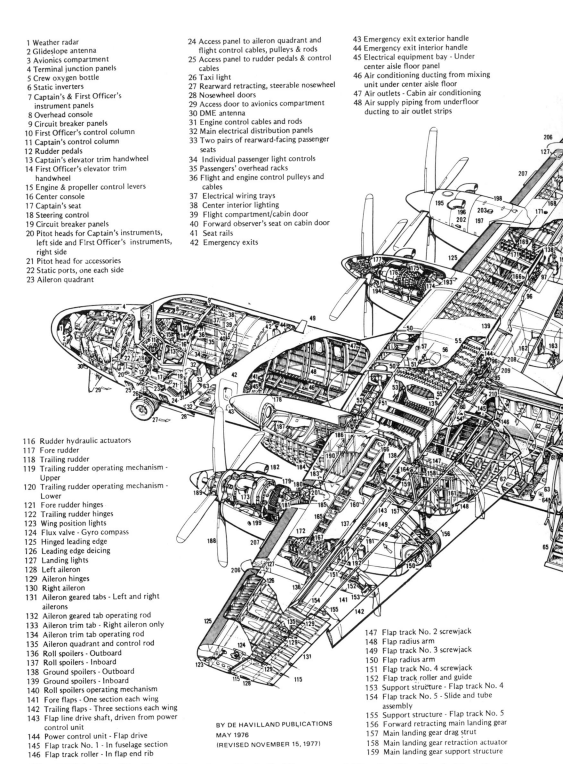

1 Weather radar
2 Glideslope antenna
3 Avionics compartment
4 Terminal junction panels
5 Crew oxygen bottle
6 Static inverters
7 Captain's & First Officer's
  instrument panels
8 Overhead console
9 Circuit breaker panels
10 First Officer's control column
11 Captain's control column
12 Rudder pedals
13 Captain's elevator trim handwheel
14 First Officer's elevator trim
   handwheel
15 Engine & propeller control levers
16 Center console
17 Captain's seat
18 Steering control
19 Circuit breaker panels
20 Pitot heads for Captain's instruments,
   left side and First Officer's instruments,
   right side
21 Pitot head for accessories
22 Static ports, one each side
23 Aileron quadrant

24 Access panel to aileron quadrant and
   flight control cables, pulleys & rods
25 Access panel to rudder pedals & control
   cables
26 Taxi light
27 Rearward retracting, steerable nosewheel
28 Nosewheel doors
29 Access door to avionics compartment
30 DME antenna
31 Engine control cables and rods
32 Main electrical distribution panels
33 Two pairs of rearward-facing passenger
   seats
34 Individual passenger light controls
35 Passengers' overhead racks
36 Flight and engine control pulleys and
   cables
37 Electrical wiring trays
38 Center interior lighting
39 Flight compartment/cabin door
40 Forward observer's seat on cabin door
41 Seat rails
42 Emergency exits

43 Emergency exit exterior handle
44 Emergency exit interior handle
45 Electrical equipment bay - Under
   center aisle floor panel
46 Air conditioning ducting from mixing
   unit under center aisle floor
47 Air outlets - Cabin air conditioning
48 Air supply piping from underfloor
   ducting to air outlet strips

116 Rudder hydraulic actuators
117 Fore rudder
118 Trailing rudder
119 Trailing rudder operating mechanism -
   Upper
120 Trailing rudder operating mechanism -
   Lower
121 Fore rudder hinges
122 Trailing rudder hinges
123 Wing position lights
124 Flux valve - Gyro compass
125 Hinged leading edge
126 Leading edge deicing
127 Landing lights
128 Left aileron
129 Aileron hinges
130 Right aileron
131 Aileron geared tabs - Left and right
   ailerons
132 Aileron geared tab operating rod
133 Aileron trim tab - Right aileron only
134 Aileron trim tab operating rod
135 Aileron quadrant and control rod
136 Roll spoilers - Outboard
137 Roll spoilers - Inboard
138 Ground spoilers - Outboard
139 Ground spoilers - Inboard
140 Roll spoilers operating mechanism
141 Fore flaps - One section each wing
142 Trailing flaps - Three sections each wing
143 Flap line drive shaft, driven from power
   control unit
144 Power control unit - Flap drive
145 Flap track No. 1 - In fuselage section
146 Flap track roller - In flap end rib

147 Flap track No. 2 screwjack
148 Flap radius arm
149 Flap track No. 3 screwjack
150 Flap radius arm
151 Flap track No. 4 screwjack
152 Flap track roller and guide
153 Support structure - Flap track No. 4
154 Flap track No. 5 - Slide and tube
   assembly
155 Support structure - Flap track No. 5
156 Forward retracting main landing gear
157 Main landing gear drag strut
158 Main landing gear retraction actuator
159 Main landing gear support structure

BY DE HAVILLAND PUBLICATIONS
MAY 1976
(REVISED NOVEMBER 15, 1977)

Fig. 2.8(b)   De Havilland Canada – Dash 7 (*Courtesy of The De Havilland Aircraft of Canada Ltd*)

49 VHF antenna
50 Water separators - Air conditioning
51 Ducting from water separators to air
   recirculating/mixing unit
52 Air cycle machine - Supplying cooling
   air to air conditioning system
53 Front spar frame - Wing/fuselage
   attachment
54 Rear spar frame - Wing/fuselage
   attachment
55 Center wing joint straps
56 Center wing access panel
57 Flight spoiler control cables and pulleys
58 Flight spoiler control quadrant
59 Aileron control quadrant
60 Aileron control cables and pulleys
61 Aileron control cables and pulleys in
   cabin bulkhead
62 Twenty-three pairs of forward-facing
   passenger seats - 50 seat configuration
63 Emergency exterior exit lights
64 Airstair door external release handle
65 Airstair door - Manually-operated,
   counter-balanced
66 Airstair door upper section
67 Cabin attendant's folding seat
68 Passenger rear emergency exit
69 Toilet compartment
70 Toilet compartment door - Sliding
71 Toilet
72 Washbowl
73 Cabin/baggage compartment door with opt-
   ional folding seat for second cabin attendant
74 Buffet unit
75 Baggage compartment - 240 cu ft
76 Baggage tiedowns
77 Baggage compartment door
78 Center fuselage/rear fuselage joint plates

79 Rear pressure bulkhead
80 Pressurization control valves
81 Rear fuselage/vertical stabilizer spar
   frame - Front
82 Rear fuselage/vertical stabilizer spar
   frame - Center
83 Rear fuselage/vertical stabilizer spar
   frame - Rear
84 Fuel filler - Pressure refuel/defuel
85 Refuel/defuel line - Rear fuselage
86 Elevator control cables and pulleys
87 Elevator trim cables and pulleys
88 Electrical wiring and conduit
89 Pressure regulator - Rudder hydraulics
90 Water separator - Deicing
91 Deicing distributor valve
92 Emergency location transmitter
   antenna
93 Refuel/defuel line - Dorsal fin
94 Deicing lines and elevator trim cables, etc
95 Master fueling valve and line vent valve

96 Fuel lines to wing tanks
97 Fuel flow control valves
98 Deicing lines
99 Rear fuselage access panel
100 Rear fuselage vent
101 Tail lower position light
102 Tail upper position light
103 Anti-collision light - Upper
104 VOR antenna
105 Horizontal stabilizer attachment points
106 Elevator control rods and cables
107 Elevator trim cables and pulleys
108 Elevator spring tab mechanism
109 Elevator spring tabs
110 Elevator trim tabs
111 Elevator trim actuator
112 Elevator hinges
113 Elevator tab horn balances
114 Left and right elevators
115 Static discharge wicks

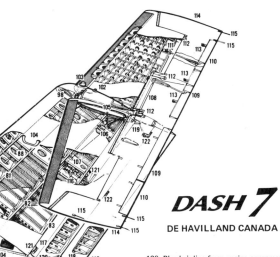

DASH *7*

DE HAVILLAND CANADA

160 Main landing gear doors
161 28 VDC, 13 amp/hr Nicad battery
162 28 VDC, 40 amp/hr Nicad battery
163 AC and DC external power receptacles -
   Under No. 3 nacelle
164 Electrical contactor boxes - One each for
   AC and DC in No. 2 and No. 3 nacelles
165 Fuel tank - Outboard, 425 U.S. (354 Imp.)
   gallons capacity
166 Fuel tank - Inboard, 315 U.S. (262 Imp.)
   gallons capacity
167 Capacitance probe - Fuel contents, 5 per
   tank
168 Access panel to outboard fuel tank
169 Access panel to inboard fuel tank
170 Access panel - Inspection of end fuel
   tank rib, etc
171 Overwing fuel fillers
172 Wing outboard joint plate
173 Pratt & Whitney PT6A-50, 1120 SHP
   free-turbine powerplant

174 Single-stage centrifugal compressor
175 Power turbine and compressor turbine
   stages
176 Engine air exhaust assembly
177 Propeller reduction gearbox
178 Engine mounting frame - Forward
179 Engine mounting pad - Aft
180 Engine mounting pad support structure
181 Engine oil filler
182 Engine air exhaust nozzles
183 Air intake deflector
184 Oil cooler
185 Access door to air intake deflector & oil
   cooler
186 Engine firewall
187 Engine air intake - de-iced
188 Hamilton Standard 24PF-305, Fully-
   feathering, constant speed, counterweight
   propeller - de-iced
189 Propeller pitch control mechanism and
   counterweights

190 Bleed air line from engine compressor to
   air cycle machine
191 Hydraulic reservoir - System is fed from
   an engine-driven, single-stage, 3000 psi
   hydraulic pump
192 Hydraulic system ground servicing panel
193 28 VDC starter/generator - One each
   engine
194 115/200 VAC generator - One each
   engine
195 Engine cowling - Forward - All nacelles
196 Engine cowling - Center - All nacelles
197 Engine cowling - Left rear - All nacelles
198 Engine cowling - Right rear - All nacelles
199 Wing inspection light
200 Tail bumper - Retractable
201 Exhaust air duct - Starter/generator
202 Air intake to engine bay
203 Exhaust air duct - From engine bay
204 Cockpit voice recorder
205 Flight data recorder
206 Wing fence
207 Wing stall bar
208 Duplicate flap drive shaft
209 Duplicate drive shaft load senser
210 Flap overspeed brake

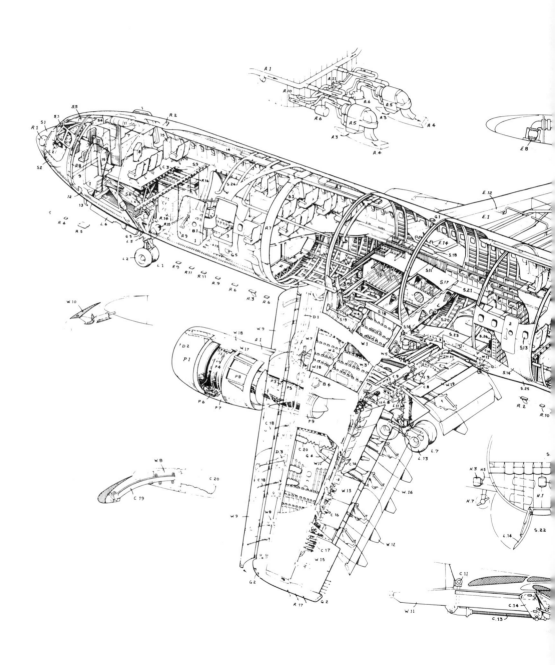

Fig. 2.9 Structural detail, Airbus A300B (*Courtesy of Airbus Industry S.A.*)

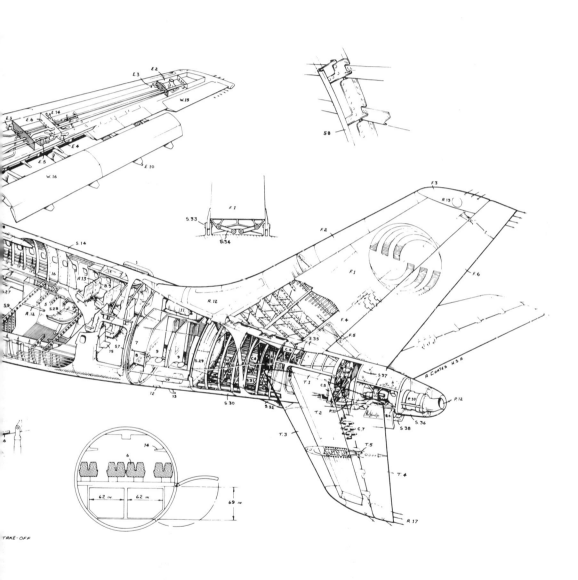

TAKE-OFF

NDING

NO STRINGERS
NO CLEATS

20

Fig. 2.10 The British Aerospace 146 (*Courtesy of British Aerospace*)

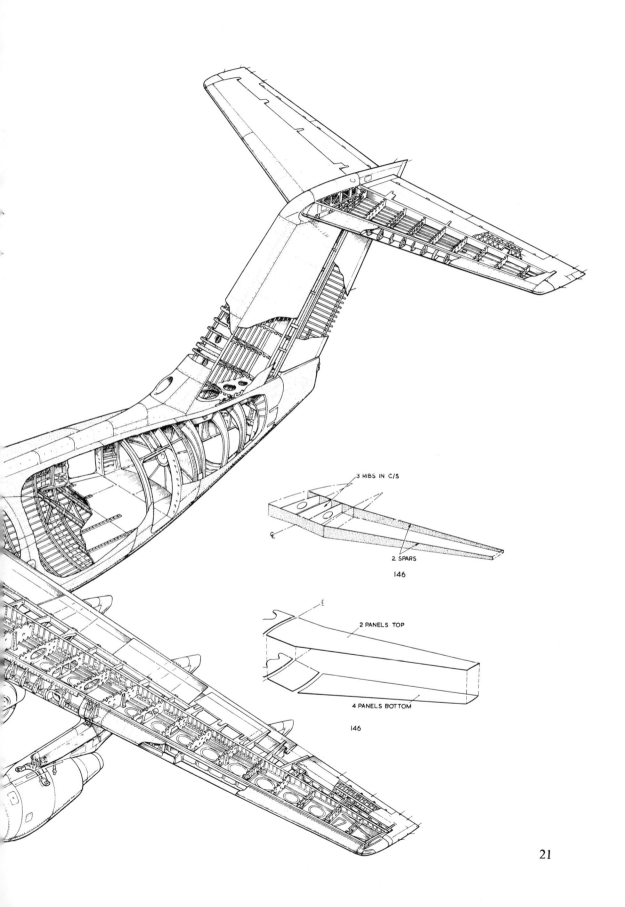

3 RIBS IN C/S

2 SPARS

146

2 PANELS TOP

4 PANELS BOTTOM

146

21

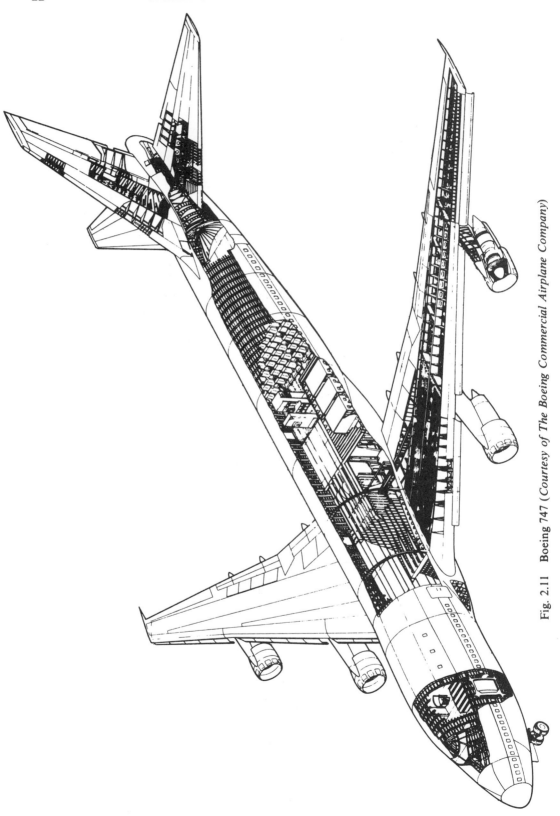

Fig. 2.11 Boeing 747 (Courtesy of The Boeing Commercial Airplane Company)

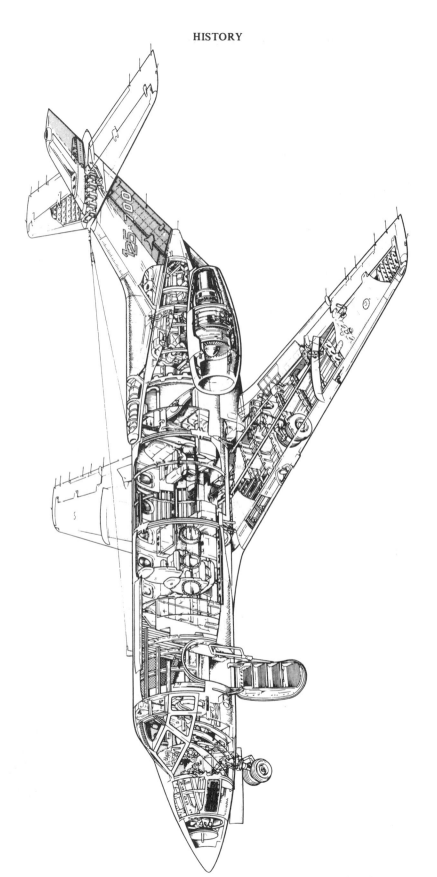

Fig. 2.12 HS 125 (Courtesy of British Aerospace)

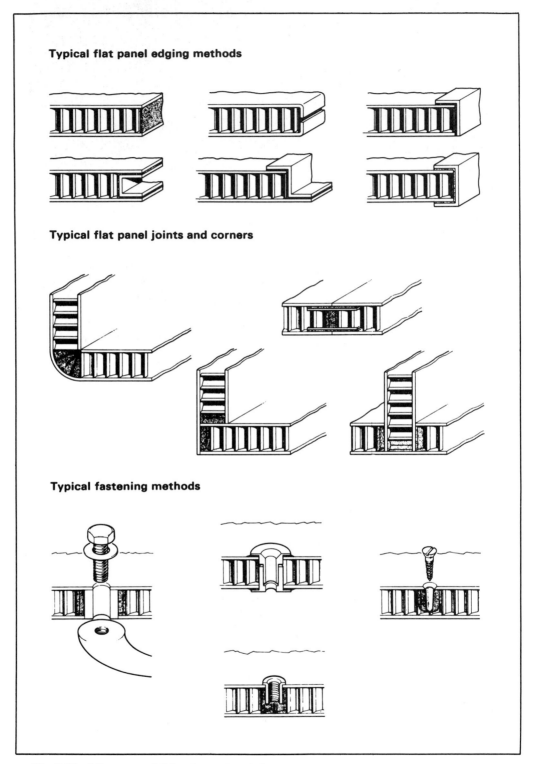

Fig. 2.13   (*Courtesy of Ciba-Geigy, Bonded Structures Division, Cambridge*)

Fig. 2.14(a)

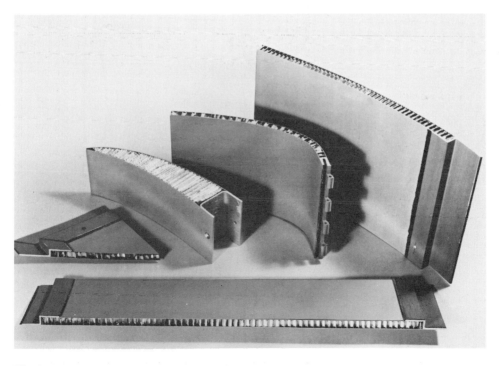

Fig. 2.14(b)

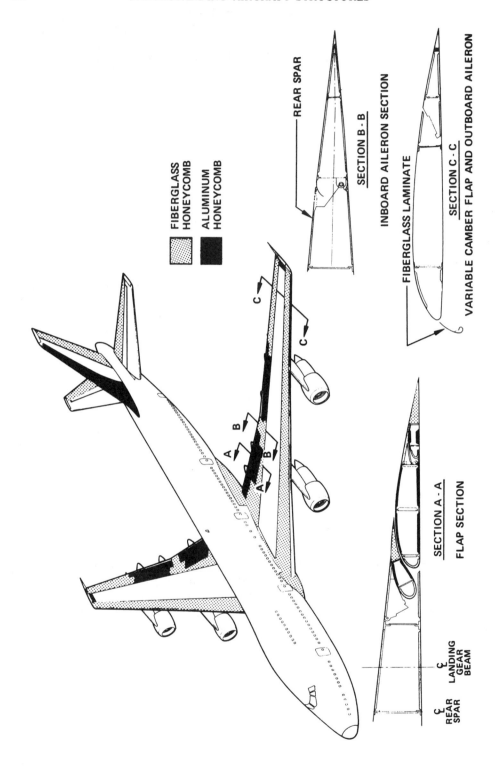

Fig. 2.14(c)   Typical use of sandwich structure ((a and (b) *Courtesy of Ciba-Geigy, Bonded Structures Division, Cambridge* (c) *Courtesy of the Boeing Commercial Airplane Company)*

Parts of the Aeroplane

3.1 The Reasons for This Chapter

3.1.1

There are many words used to describe parts of the aircraft and its operation which, although not directly the concern of the structures engineer, will be used in the text, and therefore must be clearly in mind before we start. Later chapters will examine and define terms and expressions which are particularly relevant to aircraft structures.

3.2 Terms Connected with Flight

3.2.1

The study of how an aircraft flies is called the science of Aerodynamics and the engineers who practise it are aerodynamicists. Immediately we have used a word which is mildly controversial. We used the term *aircraft* and will continue to do so throughout this book but airplane or aeroplane are equally suitable names for the whole vehicle and the reader should use the term which is in general use in his own company. Sailplanes and gliders are unpowered but so far as their structures are concerned they are aircraft in the same sense as the powered versions.

3.2.2

Aircraft engines or power plants are internal combustion engines, that is the basic fuel is burnt inside the engine and not externally as it would be in the case of a steam engine with an external boiler. The fuel is either gasolene (Avgas) or kerosene (Avtur). Power plants are either piston engines which drive propellers, or turbines which operate in one of the three following ways. A pure jet turbine is a producer of high pressure gas which is expelled backwards to thrust the aircraft forwards. Prop-jets are turbine engines driving propellers of the same pattern as those driven by piston engines, and by-pass engines are turbines which combine a pure jet function with driving a multibladed propeller enclosed in a large circular duct.

3.2.3

Aircraft are kept up in the air and can fly because of lift produced by the wing or *mainplane*. We deal with lift in more detail in chapter 4 but for the moment we will accept that a flat strip of rigid material pushed edgeways through the air will generate lift if its forward or leading edge is slightly higher than its rearwards or trailing edge.

Provided that the angle of the plate to the airflow is kept below about 10–15° the greater the angle becomes, the higher the lift becomes. Also, if the angle is kept constant but the speed through the air is increased, then more lift is generated. (Some simple but quite informative experiments on lift and other properties of flat plates or model wings can be made by a passenger in a motor vehicle holding the test piece out of a window, but the time and place for such activities should be selected with some care.)

3.2.4

If we imagine an aircraft flying straight and level at a constant speed the lift produced by (or generated by) the wing must equal the weight of the aircraft. If the lift exceeds the weight, the aircraft will be pushed higher and if the weight exceeds the lift the aircraft will sink. As the speed is constant the lift is maintained at the correct amount by keeping the angle of the wing to the airflow at a constant correct amount. In conventional aircraft this is achieved by the action of the *tailplane*. (Note here that mainplanes can be called wings but the horizontal tail surface, although very similar in shape and construction to a wing, is never referred to as a tail wing.) Other names for the tailplane are stabiliser, horizontal stabiliser or horizontal tail and it works in the following way. If the mainplane angle increases slightly the body of the aircraft is rotated in a 'tail down – nose up' direction and the tailplane is also given an increased angle to the airflow. This increased angle produces a lift on the tail which levers the body and with it the mainplane back to their original angles. A nose down attitude is similarly corrected by the tail working the opposite way and pushing the rear of the aircraft down.

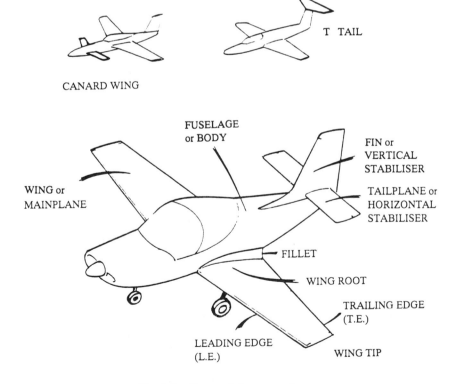

Fig. 3.1   Parts of the monoplane

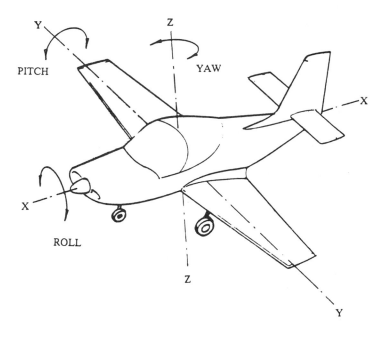

Fig. 3.2 Aeroplane axes

This tailplane action is continuous and produces the quality called stability. The degree (or quantity) of stability is decided by the designer when he considers the specification of the whole aircraft. Clearly a single seat aircraft intended for competition aerobatics and quick manoeuvres does not want to be too stable, and equally clearly a private aircraft intended for social air touring without the aid of expensive automatic pilot equipment needs to be stable so that the human pilot is not too busy just maintaining level flight. Achieving the correct amount of stability is a difficult problem for the aerodynamicist.

3.2.5

We have already seen that some parts of the aircraft can have more than one acceptable name and fig. 3.1 summarises the situation so far. This figure points out an interesting piece of aeronautical use of language. A bird has two wings but a conventional aeroplane is a monoplane, that is it has one wing. On the other hand, in the assembly shop of an aircraft manufacturing plant when one side of a wing is being attached to a fuselage the action will be called 'bolting on a wing' or something similar, it will not be 'bolting on half a wing'. Some English names of aircraft parts have a French origin reflecting the heavy French influence on early aeronautical research. In some books, although the use of the name has declined, the whole tail unit, that is the extreme rear fuselage, the vertical stabiliser and the horizontal stabiliser together with the control surfaces (see below), are referred to as the empennage. We shall notice other words with a French influence later.

3.3 Terms Connected with Control

3.3.1

When the aircraft is flying it needs to be controlled by the pilot and we think of it being controlled about three axes of movement (fig. 3.2).

One axis is an imaginary pivot line stretching from wing tip to wing tip and rotation about this axis which causes 'nose up' or 'nose down' movement is called *pitching*. In engineering terms the pitching axis is also called the YY axis.

The imaginary horizontal line extending from the extreme nose of the aircraft to the extreme tail is called the rolling or XX axis and movement about this axis is called *rolling*.

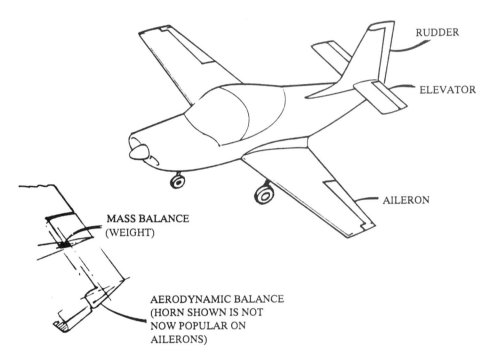

RUDDER

ELEVATOR

AILERON

MASS BALANCE
(WEIGHT)

AERODYNAMIC BALANCE
(HORN SHOWN IS NOT
NOW POPULAR ON
AILERONS)

HIGH LIFT DEVICES

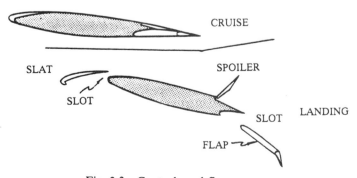

CRUISE

SLAT

SLOT

SPOILER

SLOT

LANDING

FLAP

Fig. 3.3   Controls and flaps

The third axis is vertical and passes through the intersection of the other two. Rotation about this ZZ axis is called *yaw* and hence ZZ is the yaw axis.

3.3.2

Fig. 3.3 shows the controls which effect the movements about the three axes. Pitching is controlled by the *elevators*. Roll is controlled by the *ailerons*. Yaw is controlled by the *rudder*. These three are called *control surfaces* and are usually balanced as shown in the figure. Aerodynamic balance reduces the effort required to move the surface and the mass (or static) balance avoids *flutter* which is discussed in chapter 4.

3.3.3

A further small control surface is attached to the trailing edge of the main control surface. These small surfaces are called *tabs*. The operation and purpose of trim tabs and servotabs can be investigated in any book dealing specifically with aerodynamics.

3.4 Terms Connected with High Lift Devices

3.4.1

The majority of aircraft are equipped with additional surfaces on their wings to improve lift at slow speed. Fig. 3.3 shows and names some typical high lift devices

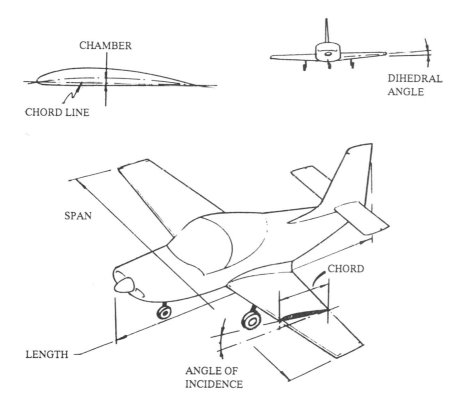

Fig. 3.4   Dimensions of aircraft

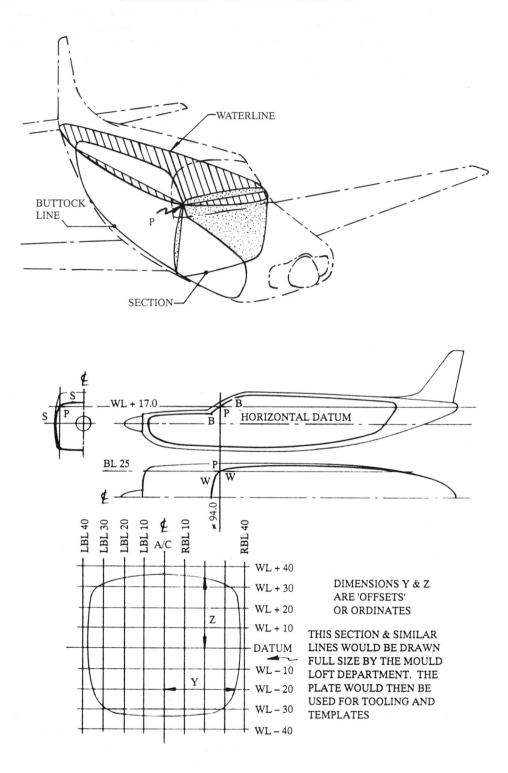

DIMENSIONS Y & Z
ARE 'OFFSETS'
OR ORDINATES

THIS SECTION & SIMILAR
LINES WOULD BE DRAWN
FULL SIZE BY THE MOULD
LOFT DEPARTMENT. THE
PLATE WOULD THEN BE
USED FOR TOOLING AND
TEMPLATES

Fig. 3.5   Lines drawing

which go under the general group name of *flaps*. Again it would be better for the reader who is interested in aerodynamics to investigate the function and working of flaps for himself. So far as this book is concerned it is sufficient to say that *slots*, which are the gaps between the flaps and the main wing, smooth out airflow and delay the stall (see also para. 4.6.3) and flaps either increase camber (see fig. 3.4), or increase wing area, or both.

3.5 Terms Associated with the Shape and Dimensions of the Aircraft

3.5.1

The names of the leading dimensions are shown on fig. 3.4. The usual practice is to quote leading dimensions in feet and inches but other dimensions, even large distances such as door openings, in inches. Parts of an inch are quoted as decimal. Thus the distance from the main undercarriage to the nose wheel on the Boeing 747 is quoted as 78′ 11.5″ (read, seventy-eight feet eleven point five inches) and the basic wing chord as 449.68 (read, four, four, nine point six eight). (Note that in the second case 'inches' can be used or implied and that the figures after the decimal point are never quoted as '. . . point sixty-eight'.) Metric dimensioning has not yet been widely adopted by the dominant American aircraft manufacturers (1991).

3.5.2

Fig. 3.5 shows the method of describing by drawing the complex shapes of an aircraft. The technique was inherited from the boat building industry which accounts for the name *waterline* for any horizontal section or cut. In the diagram the three sectional cuts are each shown as a single division, but in fact a large number of cuts would be required to describe a shape accurately. Any point on the surface of a solid shape will have a *waterline*, a *section line* and a *buttock line* passing through it. Point P on the diagram is an example. The problem for the engineer specifying the shape is that the points either side of P, that is points WW, BB and SS, must all make a smooth or fair curve with P and at the same time agree with another set of waterlines, butts and sections of their own. The process of achieving this agreement is usually computerised but in the past some very elegant aircraft have been shaped by having their 'lines' drawn out full size by the mould loft department. Each of the cuts illustrated is crossed by the other two cuts which show as lines, and if the various divisions are made at regular intervals the shapes will each be covered by a grid pattern. In fig. 3.5 the section illustrated is crossed by water and buttock lines which can then be dimensioned to define the section outline.

Sections are taken on station lines which are labelled by their distance from the aircraft datum, so that station 459.50 (this can be written Stn. 459.50 or Sta. 459.50 or # 459.50) would be 459.50 in. from the datum which is normally at the aircraft's extreme nose.

Waterlines are similarly labelled, so the WL –20.75 would be 20.75 in. below (by the minus sign) the aircraft's horizontal datum. This datum is an arbitrarily positioned line; in the case of small piston engined aircraft it is often on the centre of the propeller shaft and on large aircraft with circular section fuselages it is usually a line joining the centres of the circles in the main part of the fuselage; but note here that at nose and tail the centre of the fuselage sections would be below or above the horizontal datum.

Buttock lines are labelled by their distance from the fuselage centre line, so that RBL 11.30 would be a butt taken 11.30 in. to the right (starboard side) of the aircraft centre line.

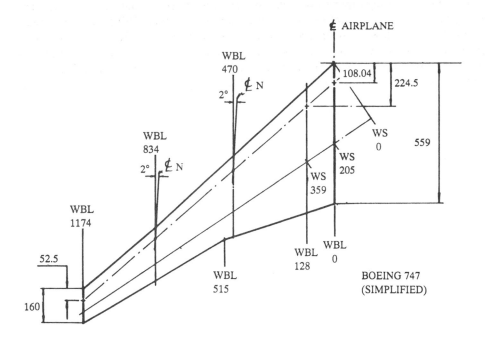

BOEING 747
(SIMPLIFIED)

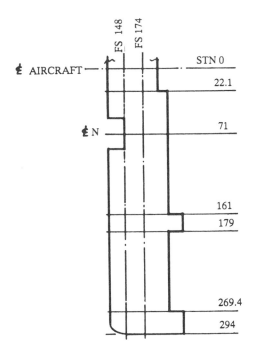

WBL WING BUTT LINE
WS WING STATION
STN STATION
FS FUSELAGE STATION
℄ N CENTRE LINE of ENGINE NACELLE

PILATUS BRITTEN-NORMAN ISLANDER
(SIMPLIFIED)

Fig. 3.6 Alternative methods of describing wing geometry

In general, fuselage frames (see chapter 2) lie on stations, but there are no other fuselage structure parts that exactly follow the major loft lines. Wing ribs are correctly defined by butt lines, but unfortunately some companies confuse the situation by labelling wing rib positions as stations. Alternative methods of describing wing geometry are illustrated in fig. 3.6; neither is more correct than the other but the differences are often confusing to the student trying to memorise and understand a mass of new words.

CHAPTER 4

Loads on the Aeroplane

Before considering the loads imposed on the structure of the aircraft, it will be as well if we remind ourselves of some of the engineering concepts and the associated terminology.

4.1 General Flight Forces

4.1.1

The method by which an aircraft flies is well known: air passing over the wing lifts by suction which is a pressure reduction; this pressure reduction over the top of the wing is assisted by an upward pressure under the wing. However, although the aircraft flies by passing through the air, it is convenient for the engineer to consider that the aircraft is standing still and that the air is moving past it. These ideas are illustrated in fig. 4.1. Fig. 4.1(b) shows a piece of paper lifting when air is blown over it. Fig. 4.1(c) is a diagram of a wing section with airflow represented by lines. Note the direction of the airflow directly behind the wing. Fig. 4.1(d) shows diagrammatically a slightly different concept of lift on a wing which is more convenient for the structures engineer. The lines are not now meant to represent airflow, they represent direction of air pressure onto the wing. The line of pressure or force behind the wing is in the same direction as the deflected airflow in fig. 4.1(c). Because the airflow has been deflected from its straight path, a force must have been applied to make it change direction. In the diagram this force is represented by the line R. From the principles shown in fig. 4.1(d) we can divide the applied force R into two parts, a vertical part which we call 'W' and a horizontal part which we call 'T'. 'W' is provided for us by the weight of the aircraft and 'T' by the thrust of the engine. To further clarify the situation we dispense with the original lines of air forces and replace them with forces more directly opposing W and T. These forces we call *lift* and *drag* and we label them L and D. Lift opposed by weight, and drag opposed by thrust, represent the balanced forces which are established by the change in direction of the airflow.

4.1.2

This concept of dividing forces into their component parts as we divided 'R' into the components 'W' and 'T' is most important and basic for students who wish to go on to a serious understanding of the methods of engineering associated with the design of aircraft structures. Fig. 4.3 illustrates some of these ways of dividing and joining forces; people not familiar with the principles should study this with some care. Before

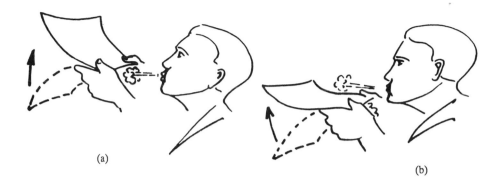

(a) BLOWING UNDER or (b) BLOWING OVER
A STRIP OF PAPER. BOTH HAVE
THE SAME EFFECT – THE PAPER LIFTS

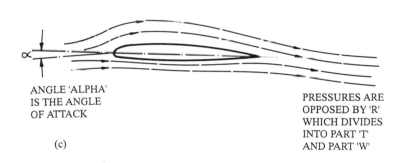

ANGLE 'ALPHA'
IS THE ANGLE
OF ATTACK

PRESSURES ARE
OPPOSED BY 'R'
WHICH DIVIDES
INTO PART 'T'
AND PART 'W'

(c)

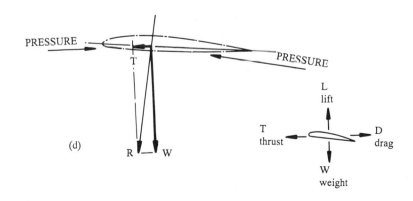

Fig. 4.1

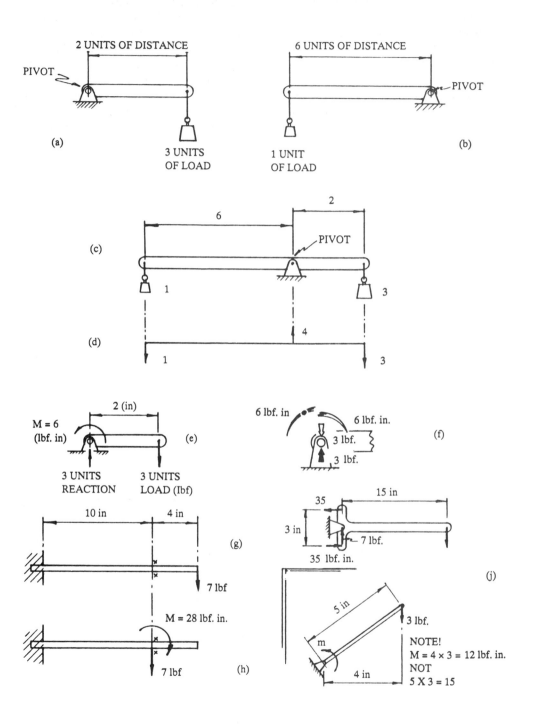

Fig. 4.2   Moments

proceeding any further, we should also remind ourselves of some of the principles associated with turning moments, balance and the transfer of loads.

Start by looking at (a) and (b) in fig. 4.2. Each of these diagrams is equally ridiculous. To stop the levers moving we would have to do something at the pivots. Joining the two bars together as shown in (c) makes the situation more possible; the lever does not swing, it is 'balanced', but the picture is still not complete. It is only completed by a vertical reaction at the pivot as shown in (d). Four units of upwards acting force are essential to oppose the four units acting downwards. In fact the downward forces cannot exist without the opposition (or reaction).

If we now redraw (a) in the way shown at (e) we can make it complete by adding a moment and a reaction.

$$
\begin{aligned}
\text{The moment (M)} &= \text{load} \times \text{distance} \\
&= 3 \times 2 \\
&= 6 \\
\text{The reaction} \quad &= \text{the load} \\
&= 3 \text{ (but note the direction)}
\end{aligned}
$$

In fig. (c) the lever is balanced because the moments each side of the pivot are equal, and the reaction of four units, is the three units in fig. (e) plus one unit from the lever to the left of the pivot.

Looking at the pivot in (e) on its own as drawn at (f), the system of forces which is acting at the pivot is as shown. The applied load and moment are the white arrows and the reacting force and moment are the black arrows. (Note here that between (e) and (f) the system has been labelled in lbf and lbf in. These read as 'pounds force' and 'pounds force inches', or just 'pound inches'. Students more used to other units may wish to relabel the diagrams so that lbf becomes N (newton) and lbf in becomes Nm (newton metre).)

Note also that the two white arrows *are* the 3 lbf load which is away at the right-hand end of the lever, and this illustrates the important principle that a load can be transferred (relocated) but in its new position it must be accompanied by a moment.

To further illustrate this last point, consider the beam in the two diagrams (g) and (h). The reactions at the wall have been omitted to simplify the drawing. Elements of the beam to the left of the plane of section xx will 'feel' exactly the same load in each case.

An engineer examining the strength of the beam at section xx would call 28 lbf in. the *bending moment* (M_{xx}) and 7 lbf the *shear* (S_{xx}).

The last diagram (j) illustrates one possible way of applying a moment at the end of a lever to balance a load. This method is used later in this chapter, see fig. 4.4.

4.1.3

Returning to our consideration of how the aeroplane flies, we have a wing with lift and drag which we attach to a body as in fig. 4.4. We said earlier that the lift balances the weight of the aircraft but it is not sensible to assume that these two forces will be directly opposing one another; without any doubt they will at times be offset from one another as shown in the diagram. This immediately produces an out of balance moment which we attend to by the load on the tailplane. By the same argument, thrust and drag are not immediately opposite one another and we also balance those out by the tailplane. All the forces on the aircraft that we have considered so far come

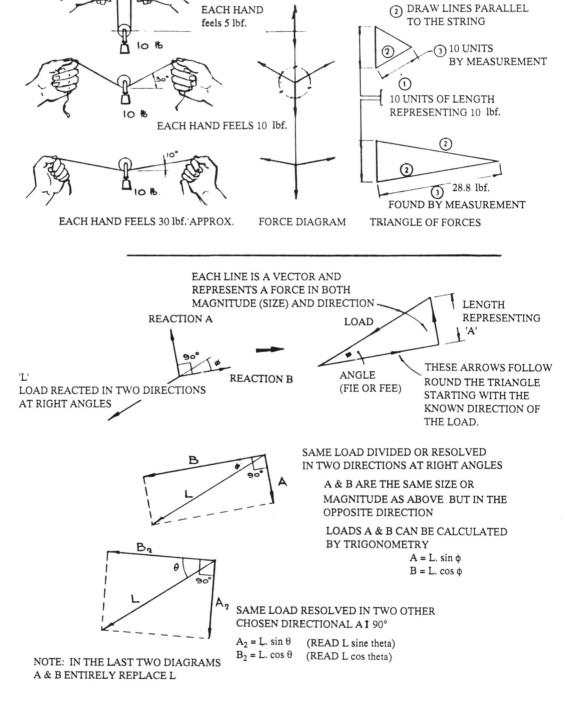

EACH HAND
feels 5 lbf.

10 lb

EACH HAND FEELS 10 lbf.

10 lb

EACH HAND FEELS 30 lbf. APPROX.

FORCE DIAGRAM

② DRAW LINES PARALLEL
TO THE STRING

③ 10 UNITS
BY MEASUREMENT

① 10 UNITS OF LENGTH
REPRESENTING 10 lbf.

③ 28.8 lbf.
FOUND BY MEASUREMENT

TRIANGLE OF FORCES

EACH LINE IS A VECTOR AND
REPRESENTS A FORCE IN BOTH
MAGNITUDE (SIZE) AND DIRECTION

REACTION A

90° φ

'L'
LOAD REACTED IN TWO DIRECTIONS
AT RIGHT ANGLES

REACTION B

LOAD

LENGTH
REPRESENTING
'A'

ANGLE
(FIE OR FEE)

THESE ARROWS FOLLOW
ROUND THE TRIANGLE
STARTING WITH THE
KNOWN DIRECTION OF
THE LOAD.

B

L

90°

A

SAME LOAD DIVIDED OR RESOLVED
IN TWO DIRECTIONS AT RIGHT ANGLES

A & B ARE THE SAME SIZE OR
MAGNITUDE AS ABOVE BUT IN THE
OPPOSITE DIRECTION

LOADS A & B CAN BE CALCULATED
BY TRIGONOMETRY

$$A = L. \sin \phi$$
$$B = L. \cos \phi$$

B_2

θ

L

90°

A_2

SAME LOAD RESOLVED IN TWO OTHER
CHOSEN DIRECTIONAL AT 90°

$A_2 = L. \sin \theta$ (READ L sine theta)
$B_2 = L. \cos \theta$ (READ L cos theta)

NOTE: IN THE LAST TWO DIAGRAMS
A & B ENTIRELY REPLACE L

Fig. 4.3 Triangle of forces

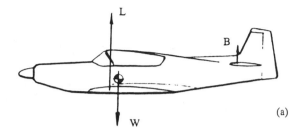

(a)

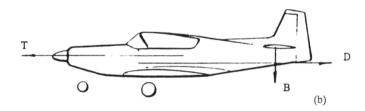

(b)

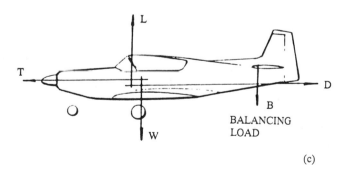

(c)

A MORE COMPLETE VERSION OF
THIS DIAGRAM IS SHOWN IN FIG. 4.8

Fig. 4.4   Flying loads

together and balance out in fig. 4.4(c). This concept of the whole aeroplane as a free stationary body surrounded by a system of balancing loads and reactions is a concept which the aircraft engineer also applies to every small part of the structure as he comes to examine it in detail.

Now consider the wing as viewed from the front. It will have lift along its whole length. For reasons of airflow, which we, as structural engineers and not aerodynamicists need only accept without question, there is more lift towards the middle of the wing than at the tips, and the lift is in fact distributed as shown in fig. 4.5. Of course this lift cannot exist on its own, it must be balanced by something, and it is balanced by the weight of the aircraft, labelled '*W*' in the diagram.

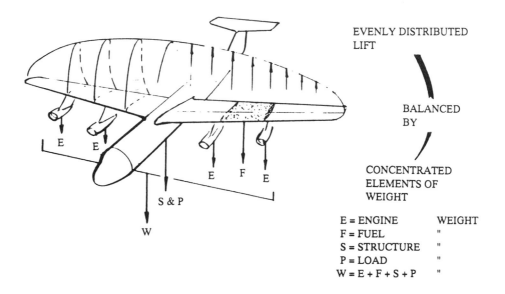

EVENLY DISTRIBUTED LIFT

BALANCED BY

CONCENTRATED ELEMENTS OF WEIGHT

E = ENGINE   WEIGHT
F = FUEL   "
S = STRUCTURE   "
P = LOAD   "
W = E + F + S + P   "

Fig. 4.5   Distribution of lift

4.2   Acceleration Loads

4.2.1

So far we have considered the aeroplane as flying straight and level. Obviously this is not always so, and since we have already said that we wish to consider it as a free body at rest under the influence of the system of balanced loads and reactions, we need a system for recognising the forces which accompany a *manoeuvre* which is the name for any disturbance of straight and level flight.

Firstly, we must be convinced that a manoeuvre always involves *acceleration*. Imagine, for instance, an aircraft flying due north which then turns 90° and heads due east. Before the turn the aircraft had no speed towards the east, after making the turn it has some speed towards the east, therefore between the two conditions it must have received an acceleration. To achieve this acceleration a force was applied to the aircraft which had to be resisted by the structure. A phenomenon which is familiar to all of us shows us a method of introducing these accelerating forces into the free stationary body idea. Imagine being in a passenger lift or elevator; just by eye we cannot tell whether or not the lift is moving. In fact, if it is moving at a steady speed, it is difficult for us to tell from the load in our legs whether the lift is going up or down. It is only when the lift stops or starts that the positive or negative acceleration appears to change our weight and the load in our legs is altered. If the lift is accelerating upwards the load in our legs is greater because the *inertia forces* in our body act in a downwards direction. (Inertia is the resistance of a body to any change in its motion. If it is standing still, it requires some external effort to move it. If it is moving, it will continue to move in a straight line and will resist the force needed to stop it or change its direction.) This is just another instance of the action and opposite reaction rule. In this case, acceleration or accelerating force acts one way and inertia the opposite way.

The analogy with the lift is easy to understand when acceleration is up and down, and because we cannot see outside the lift it is easy to accept the free stationary body idea. When the acceleration is horizontal, finding an analogy is more difficult. While travelling at a steady speed on a smooth road in a motor car, it may be difficult to admit that the loads acting on our body are exactly the same as if the car was stationary, but they are. The loads only change during acceleration away from a stop (positive or forward acceleration); during deceleration by braking (which is negative acceleration or the same as rearward acceleration); or during sideways acceleration when turning a corner. Again the 'acceleration one way, inertia the other way' rule applies, so that, if the car turns to the left, some force (the action of the front wheel tyres) has acted to the left and accelerated the car the same way, but the passenger has to restrain himself from being thrown to the right.

The essential point in the above discussion is that when the elevator or the car either stops or starts or changes direction (that is, manoeuvres), the passenger's weight appears to increase in the opposite direction to the force directing the acceleration. This is always so. Do not be confused by the apparent contradiction of this truth when the passenger lift starts to descend. The situation then is that the acceleration in the downwards direction is small and although the passenger's weight does increase in the upwards direction the increase is not enough to completely overcome his normal weight due to gravity. If the acceleration downwards became great enough, the occupant of the lift would need to push upwards on the roof to keep his feet on the floor and the greater the acceleration the harder the push would have to be. In an aircraft the same situations apply, and structural engineers deal with them by saying that during a manoeuvre the apparent weight of everything in the aircraft is increased by a factor (n) of the weight due to gravity (g). In some cases the factor n is determined by calculation but usually a figure is specified by government legislation working through its own Airworthiness Authority.

4.2.2

The presentation of the airworthiness requirement for the inertia factor or load factor is usually by the *V–n diagram* which is in a form shown in fig. 4.6. Working from the (European) Joint Airworthiness Requirements (JAR), paragraph JAR 25.337, for aircraft weighing 50 000 lb or more, n, is 2.5 and for aircraft of 14 000 lb, for instance, n is 3.1. The diagram, fig. 4.6, describes a *manoeuvring envelope* and particular combinations of speed and load within the envelope are called Cases. A similar *V–n* diagram, which describes a gust envelope, is used to present Airworthiness Requirements of the effect of air currents likely to be met by the aircraft. The air currents, or gusts, have an accelerating effect and increase apparent weights in exactly the same way as ordinary

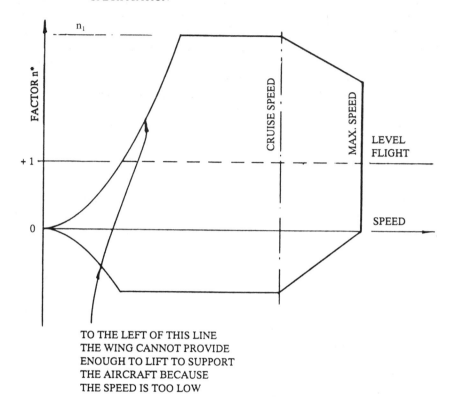

THE SHAPE OF THIS DIAGRAM
IS DERIVED FORM AIRWORTHINESS
REQUIREMENTS & THE AIRCRAFT
SPECIFICATION

TO THE LEFT OF THIS LINE
THE WING CANNOT PROVIDE
ENOUGH TO LIFT TO SUPPORT
THE AIRCRAFT BECAUSE
THE SPEED IS TOO LOW

*'n' IS THE INERTIA or
MANOEUVRING or ACCELERATION
or 'g' FACTOR.

Fig. 4.6   *V–n* diagram (the manoeuvring envelope)

manoeuvres; they may also operate in any direction and in general they become more
and more severe with higher speeds.

4.2.3

Another group of load factors specified by Airworthiness Authorities are the crash
loads or *emergency alighting loads*. In the current edition of British Civil
Airworthiness Requirements (BCAR), these are specified as 9 times *g* forwards to 1.5*g*
rearwards, 4.5*g* downwards to 2*g* upwards and 0–2.25*g* sideways. Helicopters are
shown to have rather lower factors. This method of specifying the loads along three
axes mutually at right angles raises a further interesting point. Inertias are not always
so conveniently applied and the chapter of BCAR which deals with emergency

alighting states that the prescribed loads should be taken in combination, but that the combination need not exceed 9g. It is sometimes asked why a particular figure (such as 9g) is set, the arguments being that the normal passenger travelling in an aircraft would be unable to withstand the effects of a 9g deceleration and therefore the requirement should be set at a lower figure; or alternatively that by increasing the requirement to 12g and ensuring a stronger structure, more passengers would survive a crash. These arguments might well be true, but a figure must be set somewhere and the Airworthiness Authorities consider carefully the statistics available and apply their expertise before fixing a factor at a certain level. Like most aspects of aircraft design, there is an element of compromise involved. It would be easy to set a figure which produced a very strong, safe aeroplane that was too heavy to fly, and the figure chosen is a balance of careful thought with the bias, no doubt, on the side of passenger safety. Some of the arguments used in choosing a figure are set out in Tye's *Handbook of Aeronautics* (see para. 4.7).

4.2.4

In case this question of load factors is not absolutely clear we will consider the specific example of a piece of equipment bolted down to the floor of an aircraft and subjected to the emergency alighting conditions mentioned above. If the piece of equipment is a rectangular box of 10 lb weight, to satisfy the requirement of 4.5g downwards, the floor will have to be strong enough to support 45 lbf (read forty-five pounds force) pushing down, i.e. 10 lb × 4.5g. (Note that 4.5g is just a factor with no dimensions but it does alter a weight to a force.) Similarly the 9g forwards requirement means that the attachments must resist 90 lbf parallel to the floor. If we are working in SI units the arithmetic is slightly more complicated. A weight of 1 kg at an acceleration of 1g produces 1 kgf (read one kilogramme force) but since the unit of force in SI units is the newton, we must convert. The conversion factor is 1 kgf = 9.81 N.

So   7 kg at 2.25g   =   15.75 kgf
therefore 15.75 kgf × 9.81 = <u>154.5 N</u>

For a quick and rough calculation use a factor of 10 and say:

15.75 kgf = <u>157 N</u> (which is correct to less than 2%)

(Another useful conversion factor is 1 lbf = 4.45 N.)

4.3 Further Aerodynamic Loads

4.3.1

In fig. 4.4 the load positions shown are exaggerated to illustrate the idea of the aircraft in balance. If this book was written about aerodynamics we should be a lot more careful about lift and drag and how and where they act, but as we are concerned with structures a few more ideas on air loads will suffice.

Fig. 4.1(c) showed a wing section (or airfoil or aerofoil) at an angle α (read alpha) to an air stream. This angle is called the angle of attack and there is a similar angle called the angle of incidence which is measured to the horizontal datum of the aircraft (see fig. 3.5).

As the angle of attack is increased from its usual working or cruising speed attitude of about 2°, the lift and drag will increase. At about 15° the airflow will no longer follow round the upper surface, the wing will 'stall' and lift will drop to zero.

Conversely, if the angle of attack is reduced, lift will decrease to zero at a small negative angle and then start to produce the negative lift which allows aerobatic aircraft to fly upside down. While the lift and drag are changing, the point at which they act effectively, which is called the *centre of pressure* (or CP), is also changing. At the angle of zero lift it is, for some wings, right back near the trailing edge and it moves forward as the angle is increased, usually to about a third of the wing chord distance. To clarify this situation, the aerodynamicists quote the lift and drag as acting through one imaginary point for all angles of attack and add a varying moment to take care of the CP shift. This imaginary point is usually at 25% of the wing chord and is called the *quarter chord* point.

4.3.2

Lift also varies with wing area, air density and aircraft speed (in fact as speed squared, i.e. twice the speed, four times the lift). It is also affected by *aspect ratio*.

For an absolutely rectangular wing, aspect ratio (AR) = span/chord, so a short stubby wing has low aspect ratio and a long slender wing, like a sailplane, has high aspect ratio. Now if we think about the airflow past a short wing with the air pressure reducing as it passes over the wing, we can imagine this reduced pressure pulling some air in from the ends. The effect of this is to change the line of the airflow from a direct path between leading and trailing edges to a longer path slightly along the wing. With the leading edge a little higher than the trailing edge to give an apparent angle of attack, the longer route taken by the air gives a reduced real angle of attack and hence a lower lift. If wings could have an infinite span the air would travel square across the wing for greater efficiency. They cannot have infinite span but the higher the aspect ratio the better they are aerodynamically. Unfortunately, the structure is affected in just the opposite way; a short stubby wing is lighter for the same area than a long slender wing; so, as frequently happens in aircraft design, compromises have to be made in order that a particular requirement can be met.

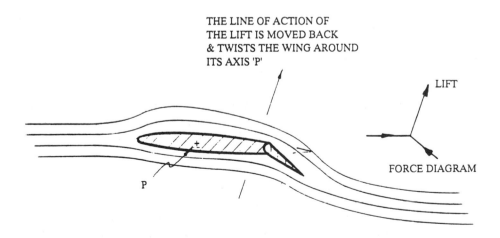

Fig. 4.7   Airflow deflected by a control surface

4.3.3

We will now consider the basic control surfaces, the ailerons, elevators and rudder, which were shown on fig. 3.3. Fig. 4.7 could be any of these surfaces, and when moved to the position shown it deflects the airflow and produces lift, drag and moment as we saw earlier in this chapter. These forces are of interest to us because the structure has to resist them, or more accurately, has to transmit these forces so that they move the mass of the aircraft. The force needed to rotate a large weighty aircraft is considerable and if we say fig. 4.7 shows the elevator of a Boeing 747, then the loads at the hinges will obviously be many thousands of pounds force. If, on the other hand, we could imagine anything so unlikely as a Boeing 747 that was as light as a balloon, or if we put our 747 size elevator on a small light aircraft, then the loads needed to rotate the aircraft would be very much smaller. In fact we could have (if our imagination were strong enough) the same size elevator, the same angle of movement, the same air speed but much smaller loads at the hinges. What we are saying is that to exist at all, a load needs to be resisted. Without the bulky, weighty inertia of a big aeroplane to push against, our hinge loads fade away. It is obvious that the structural loads imposed by control movements relate to the size and weight of the aircraft they are mounted on; it is less obvious that, for the same size aircraft, control loads are altered by the weight being lifted. Control loads also vary with the rate of response of the aircraft. For instance it takes more aileron load to roll an aircraft quickly than to roll it slowly. Methods of assessing all these loads are found in the Airworthiness Requirements publications.

4.3.4

Ailerons cause special problems: in the 1920s they effectively started a whole new science which became the study of *aeroelasticity*.

If we say that fig. 4.7 shows an aileron out on the tip of a long flexible wing, by deflecting the control down we are hoping to increase the lift at that wing tip and cause the aircraft to roll. However, because the additional lift is so far back and the wing is flexible, the whole cross-section in the diagram will be twisted anticlockwise until the resistance of the structure stops it. If the rotation is sufficient, the lift on the wing, far from being increased by the control movement, will actually be decreased causing the aircraft to roll the opposite way to that expected. David B. Thurston in his very excellent book, *Design for Flying*, (para. 4.7), says of this control reversal effect '. . . [it] could provide quite a bit of activity in the cockpit . . .'.

We took an 'aileron down' case to illustrate the effect but 'aileron up' shows a possible, even worse effect where forced rotation of the wing section causes a local stall, whereupon the unloaded wing springs back until the airflow restores itself and the process starts over again. This is one form of a phenomenon called *flutter* which can be induced in all sorts of ways. Its study is complex but well documented and modern aircraft do not suffer the catastrophic failures created by flutter in early aircraft before its various causes were recognised. The general name, aeroelasticity, is given to the study of these problems of control reversal and flutter, together with control surface balancing, resonance of vibrations and any other effect which follows from the fact that aircraft structures are springy and actually change their shape under load. Work in this area is probably the most difficult that the aircraft stressman and designer have to deal with. As we said of flutter, a lot of study has gone into the subject, especially mathematical analysis which has been made possible by the computer power now available. Unfortunately there are no 'rule of thumb' methods for the certain elimination of aeroelastic problems. The Airworthiness Authorities lay down requirements and test procedures which eliminate the possibility of danger to the airline passenger or the private pilot, but it is in the nature of aircraft design that performance and technology must be pressed forward right to their limits and that is where aeroelastic problems abound.

4.4 Other Loads

4.4.1

We have concentrated entirely on flying loads, apart from a mention of crash loads, but there are others which affect the structure. The *loads due to pressurisation* are important, especially if the fuselage is not circular. A fuselage with a non-circular cross-section will try to become circular under internal pressure and impose bending loads on the fuselage frames. The 'double bubble' style of fuselage is a clever exception, with the shape maintained by tension in the floor beams (note! the term floor 'joists' is not normally used in aircraft structure).

A tragic but instructive accident illustrated an interesting pressurisation structural problem. In this accident the lower fuselage below the floor was suddenly and totally depressurised by the opening of an unlocked cargo door during flight. The floor beams, which were parallel sided constant section bending members then collapsed. In fact they tried to adopt a shape which was part of a circle. No structures engineer would criticise another for not foreseeing this illustration of the power of pressurisation loads, which led, interestingly to the immediate adoption of automatic pressure equalising valves between compartments.

4.4.2

Landing loads also need to be considered. Airworthiness Requirements give advice and assistance with the assessment of landing loads which can have a large influence on the structure. For instance on a normally configured twin-engined aircraft in flight, the weight of the engines, the fuel, the main undercarriage and the wing itself, are all nicely spread out along the wing and directly supported by the lift. On the ground, and worse still during a heavy landing, all those same weights are propped up on the top of the undercarriage leg which is a more difficult situation for the wing structure to deal with.

4.4.3

High-speed flying produces another type of structural load which needs a classification of its own because it exists in addition to the flight and manoeuvring loads. This loading is an effect of heat produced by high-speed flight and the distortion which the heat causes.

As the aircraft forces its way through the air, parts of the structure, especially the surface skinning, are warmed and at very high speeds become quite hot. At Mach 2 (read mark two), that is at twice the speed of sound, some areas reach 150° C. This type of heating, called kinetic heating because of its association with movement, produces different temperatures on different parts of the structure. The temperature variations cause expansions which are different from member to member, and thus the internal loads that we are discussing are produced: members trying to expand because of the heat are being forcibly restrained by cooler members. (Note that both hot and cold members are loaded.) Although this idea is not easy to understand without a more general appreciation of the form of structures, it is an important source of load in the aircraft and will be referred to again in chapter 11.

4.5 Further Load Factors

Assessment of the loads on each piece of structure under investigation constitutes more than half of the stressman's job. The notes in this chapter have dealt with major loads, details of which would be supplied to the structures department by the aerodynamics and wind tunnel departments (and also, during a development programme, by the flight test department).

Once the major loads have been established the designers and stressmen have to break each load down and distribute it through the structure. For instance they might find that they cannot make two hinges each strong enough to carry half an elevator load and therefore they will need to use three hinges.

In a later chapter we shall discuss *proof* and *ultimate stress* (see para. 5.5.12) which, in brief, are, respectively, working strength and breaking strength for a material. The majority of loads derived from specifications and Airworthiness Requirements are called *limit loads* or *working loads*. To compare these with the abilities of the material they must first be multiplied by a *proof factor* to give a *proof load*. This proof factor is also specified in Airworthiness Requirements (it is usually 1.0 so that proof load is numerically the same as limit load). The proof load is a load that the structure is expected to cope with repeatedly and without distress. Crash case loads (or emergency alighting loads) are usually 'ultimate' loads, that is the structure must be strong enough to withstand the load once, but is then understood to be too stretched or distorted for further service.

The connection between proof and ultimate loads applied to a structure, and the working and breaking strengths for the material of which the structure is made, is clear and will be made clearer after reading chapter 5. There is a further connection built into the Airworthiness Requirements called an ultimate factor. This is a factor of 1.5 for most civil aircraft requirements, and works in the way shown in the following simple example.

A stressman is investigating a bracket supporting a piece of equipment weighing 10 lb. He checks with the Airworthiness Requirements and finds that (for one Case):

the limit load factor (which is quoted as a manoeuvring factor) $= 2.5g$
the proof factor $= 1.0$
and the ultimate factor $= 1.5$
therefore the proof load on the bracket $= 10\,\text{lb} \times 2.5g$
$= 25\,\text{lbf}$
and the ultimate load on the bracket $= 25\,\text{lbf} \times 1.5$
$= 37.5\,\text{lbf}$

Strictly speaking the stressman should now make two sets of calculations, one set comparing 25 lbf with the proof stress that is the *allowable* proof stress of the material of the bracket, and the other set comparing 37.5 lbf with the ultimate stress and *allowable* ultimate stress of the material. In practice he would be investigating a number of components all made of the same metal and he would know the relationship between its proof and ultimate stresses. If, for instance, he knew that the proof stress was greater than two thirds of the ultimate stress he would know that by basing his calculations on the ultimate loads and the ultimate stress he would also be covering the proof conditions.

4.6   The Whole Aircraft

With the aid of diagrams we will consider the arrangement of loads on the whole aircraft in four cases, cruising straight and level, turning, stalling and landing.

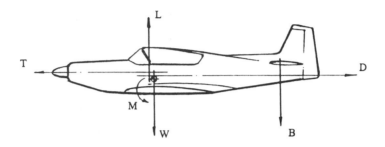

W = WEIGHT
L = LIFT (AT THE WING
        AERODYNAMIC CENTRE)
M = MOMENT (ABOUT THE
        AERODYNAMIC CENTRE)
T = THRUST
D = DRAG
B = BALANCING LOAD
        (FROM THE TAILPLANE)

NOTE THIS DIAGRAM
IS SIMILAR TO FIG 4.4
BUT SHOWS THE MOMENT
MENTIONED IN PARA 4.3.1

Fig. 4.8   Cruising flight

4.6.1

The loading situation in the straight and level cruise condition is as shown in fig. 4.8.

In paras. 4.2.1 and 4.2.2 we discussed inertia factors (load factors) which are written as *ng*. In the cruise the inertia factor is 1*g* and applies to the weights of all the elements of structure and equipment which constitute the whole aircraft. Sometimes the inertia factor is referred to as an acceleration factor or acceleration coefficient. There is nothing wrong with such a descriptive name except that in the straight and level condition we are left to explain the statement '. . . the acceleration factor is 1*g*', when clearly the aircraft is not being accelerated in any direction. Remember that whether it is called 'inertia', 'load' or 'acceleration' it is always a factor and as such is a device used by engineers to bring into calculation the apparent increase in weight of objects which are being accelerated. As a multiplying factor it represents an unaltered situation when it is 1.0.

4.6.2

When an aircraft is turning it suffers an apparent weight increase due to the centrifugal force which is trying to make it go straight on instead of round the curved path of the turn. In a correctly executed manoeuvre the loads balance as shown in fig. 4.9.

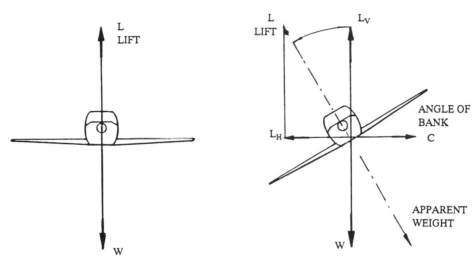

L IS RESOLVED INTO L_V & L_H

THE ANGLE OF BANK IS NEEDED TO PRODUCE A FORCE L_H TO BALANCE THE CENTRIFUGAL FORCE C.
NOTE THAT L HAS TO BE GREATER IN THE BANK

Fig. 4.9   Loads in a turn

4.6.3

Stalling occurs when the lift surfaces of the aircraft are no longer able to provide enough lift to balance the weight, and the speed of the aircraft at which this occurs is called the *stalling speed*. Even this is not simple, because we can have one stalling speed with a 'clean' wing and another lower speed when flaps, slots and any other of a selection of lift-improving devices are added. We also have the situation described in para. 4.6.2, where during a turn the apparent weight of the aircraft is increased, so that the ability of the wing to provide enough lift will disappear at a rather higher speed in a turn than when the aircraft is flying straight and level. The aerodynamic action of stalling is a breakdown of the airflow over the top surface of the wing as shown in fig. 4.1 and, except for some high-speed situations, occurs at an angle of attack of about 15°.

The structural load situation is that immediately after the stalling of the wing the aircraft starts to *free fall*, that is all the pieces of structure and equipment become apparently weightless. In para. 4.1.3 we spoke of the 'free body' concept with all the loads in balance, so we might expect that when the lift disappears the opposing and balancing weight will also disappear. In fact, the ideal situation is not quite realised, because although the main wing has stalled, the tailplane is still providing positive lift due to its attitude and angle to the airflow. This positive lift is an unbalanced load and therefore produces an acceleration. (This is similar to the situations of para. 4.2.1 when we said that where we have an acceleration we must have a force, here we have a force so we must have an acceleration which lasts until a new balance of forces is struck.) In the stall case the unbalanced tailplane load causes a nose down pitching rotation which aerodynamically allows the aircraft speed to build up and stable flying conditions to be restored (see fig. 4.10). Structurally the loads involved on the main structure in the stall are not usually of any significance but the loads on control surfaces and flaps may be important.

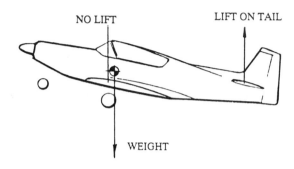

Fig. 4.10   Loads out of balance in a stall

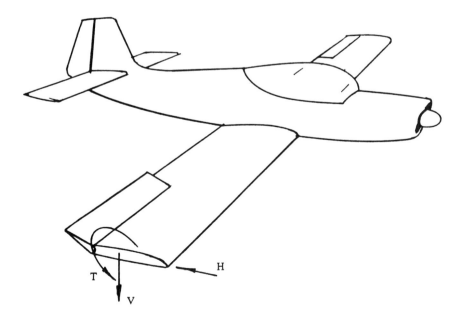

Fig. 4.11   Loads on the wing

4.6.4

Landing loads are a major consideration in the design of the structure and are involved with the designed cushioning characteristics of the undercarriage. The stiffer the springs of the undercarriage the more rapid the vertical deceleration of the landing aircraft and the higher the loads involved on the structure. The Airworthiness Requirements are very descriptive and demanding on this subject.

4.6.5

If we look back at figs. 4.7 to 4.9 we can see that the general loading on any major part of the aircraft structure is a bending or a twisting or both. This general loading pattern is shown in fig. 4.11 and will be discussed in the next chapter.

4.7 References

Thurston, D. B., *Design for Flying*, New York: McGraw-Hill Book Co. (See also books listed in para. 12.3.)

Tye, W. *Handbook of Aeronautics*, No. 1, *Structural Principles and Data*, Part 1 'Structural airworthiness', 4th edn., London: Pitman.

CHAPTER 5

The Form of Structures

5.1 Structure Relative to Aircraft Design

In the last chapter we looked at the loads which the aircraft has to carry; in this chapter we consider how the structure is arranged to support the loads (or, more exactly, how the structure links the loads to the reactions).

Perhaps the most significant thing about a piece of aircraft structure is that it is a compromise. Some years ago an amusing cartoon was printed showing how different people in the aircraft business thought of an aeroplane. To the commercial operator it was a flying box for carrying people or freight; to the avionics engineer it was just a means of getting his radar aerials and navigation aids airborne; to the aerodynamicist it was a slim, shining and beautiful creation with no unsightly access panels or skin joints or rivet heads, not even room for passengers; to the structural engineer it was a few simple robust members reminiscent of a railway bridge or a tower crane. Like all good cartoons this was not totally ridiculous. An aeroplane is all of these things but not one thing alone, so compromises have to be made. The structures engineer is constantly compromising with his own requirements, for example discarding methods or materials which would ideally satisfy one set of conditions but which would be unsuitable for another set, or correcting an apparently satisfactory design which proves to be impossible to manufacture. However, he mainly needs to adjust his ideas to satisfy both the aerodynamicist, who (broadly speaking) specifies the external shape of the aircraft, and the payload, that is the passengers, freight or armament, which together with the fuel, determine the internal dimensions.

5.2 Historic Form of Structure

In the early days of flying the aircraft structure or *airframe* had almost only one requirement, it was the minimum which the designer, who was probably also the test pilot, thought adequate to keep the wings intact and support the control surfaces and the pilot (see fig. 2.1). Later, the basic structure was covered with *fairing* which improved the streamline shape without contributing to the strength (see fig. 2.3). Wings, of course, had from the beginning been covered with fabric or paper, but this covering did not form part of the structure (except that very locally it carried air loads to the ribs). The third phase, which is the method used almost universally on modern aircraft, incorporated the fairings into the load-carrying structure or looking at it another way, pushed the structure right out so that it filled the aerodynamic shape. This type of structure, which, as we said in chapter 2, is called stressed skin or semi monocoque, is efficient in the engineering sense that it has material or components

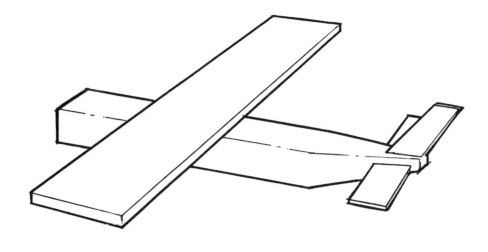

(a)

(b)

Fig. 5.1   Stuctural boxes or 'tubes'

performing more than one function. In some aircraft the wing skins perform three functions: they transmit the air loads which provide lift, they make the tube or box which is the strength of the wing (see para.5.6.8) and they are also the walls of the fuel tanks. Although it is very difficult to imagine a better type of structure for aircraft, it is always possible that one will be invented. There is even now a lower limit of size and weight of aircraft below which the stressed skin would become so thin that it would be unmanageable. Man-powered aircraft, for instance, are so light that the structure must belong to the second phase of structural development because the covering is too flimsy to be made 'load carrying'.

5.3   General Form of Structure

The aircraft structure in its various phases of development has always consisted of an assembly of large built up tubes or boxes. These tubes are most clearly obvious in the stressed skin type of structure which from now will be considered as the conventional or normal type. Fig. 5.1(a) illustrates the tubes in their simplest form, and fig. 5.1(b) shows the refinement which is the basis of the vast majority of aircraft, from single seaters to wide-bodied passenger airliners and supersonic bombers. The conventional method of construction is shown in figs. 5.2(a) and (b) with an earlier type of construction illustrated in fig. 5.3. This earlier construction does not so obviously form a tube but its construction allows it to carry the same types of load as the conventional structure. These are bending and torsion loads, and before proceeding further it would be as well to define these and some of the other terms used by structures engineers.

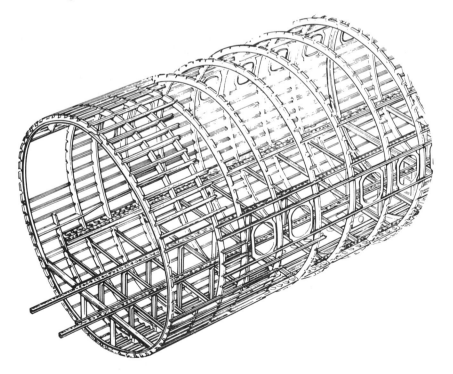

Fig. 5.2(a)   Conventional stressed skin construction – De Havilland Canada – Dash 7
(*Courtesy of The De Havilland Aircraft of Canada Ltd*)

Fig. 5.2(b)   Conventional stressed skin construction – British Aerospace 146 (*Courtesy of British Aerospace*)

5.4   Definitions of Basic Load Systems in Structures

There are three basic combinations of load and resistance to load:

shear
tension
compression

It is hard to find aircraft examples of these load/load resistance systems where the load application can quite clearly be called one of the above without any element of another. The examples in figs. 5.4 and 5.5 are close to being 'pure'.

In fig. 5.4(a) we can say that the bolt is in pure *shear* through faces AA and BB. That is, the bolt is resisting the pull by its reluctance to divide in the way shown in fig. 5.4(b). In the example illustrated in fig. 5.5 the bolt is in *tension* and will eventually divide as in fig. 5.5(b). The act of division is called a failure.

Fig. 5.3

(a)

(b)

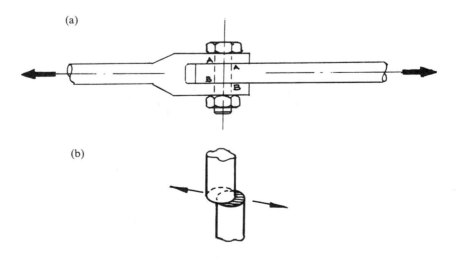

Fig. 5.4   Shear

Compression is when the load application is the opposite of tension (see fig. 5.6) and a piece of structure failing in compression is most difficult to show in a pure form because the start of a failure usually changes the form of loading. For instance, in fig. 5.6(a) one side of the round block would probably collapse a little before the other and the load would no longer be 'pure'.

(a)

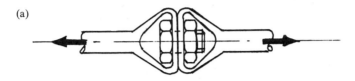

(b)

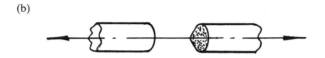

Fig. 5.5 Tension

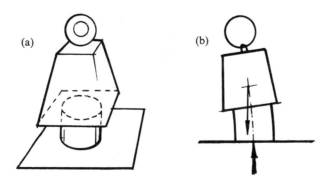

Fig. 5.6 Compression

In figs. 5.4(b), 5.5(b) and 5.6(b) load is indicated by two arrows. To finish the diagrams, we should add two more arrows to complete the whole load/load resistance system (see fig. 5.7). It is most important to note that the arrows always come in pairs, each pair being in accordance with Newton's Law which states that there cannot be a load without a reaction. Aircraft engineers tend to be rather untidy in their use of the terms but in general they would call the black arrows loads and the white arrows reactions. Unfortunately for the student of structures, he will also find the black arrows paired together as load and reaction. (The use of black and white arrows is not a convention in general use but is used just in these diagrams to illustrate the examples.)

The next two most fundamental load/reaction systems are:

torsion
bending

Torsion is very similar to shear but instead of the material being loaded as shown in fig. 5.4(a) and eventually dividing as shown in fig. 5.4(b), the torsion or twisting action is as shown in fig. 5.8(a) and the break in fig. 5.8(b) (see page 62).

Bending is the most commonly used load/reaction system in all types of structure. Fig. 5.9 shows a piece of material being bent, but the way in which it eventually breaks (called the mode of failure) is likely to be complicated as will be described in more detail in para. 5.6 below.

In figs. 5.4–5.6 (pp. 58, 59) it cannot be emphasised too much that the simple pieces of structure shown are systems complete with load forces and reaction (or resistance) forces. Without a resistance there cannot be a load. For the student of aircraft structures this rule is absolutely fundamental. Sometimes it is difficult to 'see' the two opposing forces and to keep the rule in mind, but anyone wishing to make a serious analysis of a piece of structure (especially students becoming aircraft stress engineers) will find that once they accept the truth of the rule all their problems will be greatly eased.

5.5. Definitions of Forms of Stress in Materials

Having talked about types of load systems, we will now define some of the terms used to describe materials.

5.5.1

For the aircraft engineer the words *stress* and *strain* have particular meanings which are not exactly the same as those meanings which apply in normal conversational English. From *The Concise Oxford Dictionary* we find that strain means a pull or stretching force; an engineer would not use the word in this way. To the structures engineer strain is the change in dimensions of a body brought about by force. Again we have a rather careless usage because when engineers say 'strain' they mean *unit strain* which is a ratio, being the change in length associated with a particular load divided by the original length. For example, refer to fig. 5.7(a): if the tie is loaded as shown it will stretch or extend a little. If we divide the extension by the original length we have the strain (or more exactly, the unit strain). This is clearer if we make a numerical example by giving the symbols in fig. 5.7(a) some quantities.

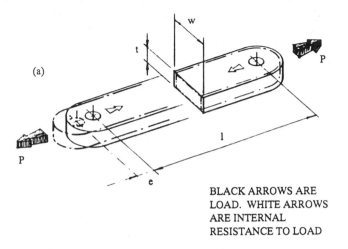

(a)

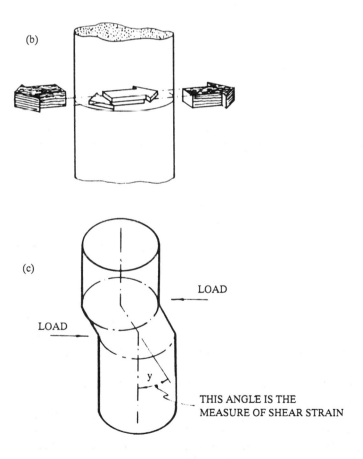

BLACK ARROWS ARE
LOAD. WHITE ARROWS
ARE INTERNAL
RESISTANCE TO LOAD

Fig. 5.7 (a)   A tension member; (b) a shear member; (c) shear strain

Numerical Examples

1. Imperial units
   Referring to fig. 5.7(a)

Length	(l)	$=$	5.0 in
Load	(P)	$=$	1000 lbf (pounds force)
Extension	(e)	$=$	0.013 in

   $$\text{STRAIN*} \quad (\epsilon) = \frac{e}{l} = \frac{0.013 \text{ in}}{5.0 \quad \text{in}}$$

   $$= \underline{0.0026}$$

2. Metric units
   Referring to fig. 5.7(a)

Length	(l)	$=$	125 mm
Load	(P)	$=$	4500 N
Extension	(e)	$=$	0.326 mm

   $$\text{STRAIN*} \quad (\epsilon) = \frac{e}{l} = \underline{0.0026}$$

Note: Strain is a ratio and has no dimensions.
* For ϵ read 'epsilon'.

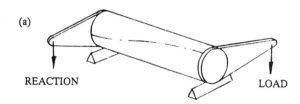

(a)

REACTION   LOAD

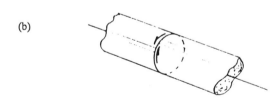

(b)

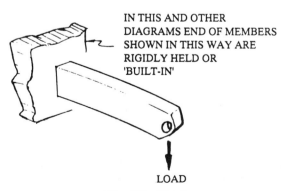

IN THIS AND OTHER
DIAGRAMS END OF MEMBERS
SHOWN IN THIS WAY ARE
RIGIDLY HELD OR
'BUILT-IN'

LOAD

Fig. 5.9   Bending

5.5.2

The stress in a member is a measure of the load compared with the amount of material available for carrying the load. Referring back to fig. 5.5(a), it is clear that if the bolt was twice as big it could either carry double the load or carry the same load twice as easily. Now, 'twice as big' does not mean twice the diameter, but twice the cross-sectional area, because two bolts in place of the original one would have the same effect. So for the engineer, stress is load per unit area (= load/area). Again this is clarified by putting some quantities into fig. 5.7(a).

Numerical Examples

1. Imperial units
   Referring to fig. 5.7(a)

Width	(W)	= 1.0 in
Thickness	(t)	= 0.036 in
Load	(P)	= 1000 lbf
Area	(A)	$= W \times t = 0.036$ in^2

   STRESS* $\quad (\sigma) = \dfrac{P}{A} \quad = 27\,778$ lbf/in^2
   (read pounds force per square inch)

2. Metric units
   Referring to fig. 5.7(a)

Width	(W)	= 25 mm
Thickness	(t)	= 1 mm
Load	(P)	= 4500 N
Area	(A)	= 25 mm^2

   STRESS* $\quad (\sigma) = \dfrac{P}{A} = 180$ N/mm^2
   (read newtons per square millimetre)

* For σ read 'sigma'.
Sometimes stress is given the symbol 'f'.

Note that we have not said what the material is but the stress is the same whether the member is made of steel or of aluminium or possibly of strong plastic, although in that case our calculations would lead us to expect that plastic would not be strong enough.

5.5.3

Paragraphs 5.5.1 and 5.5.2 above refer to tensile stress and strain, but the similarities to compressive stress and strain are quite clear.

5.5.4

In fig 5.7(b) we have a member in shear. The black arrows represent the load and the tinted area represents the material resisting the load, so again stress (in this case shear stress) is load over area and is given the symbol τ (read 'taw').

5.5.5

Shear strain is a slightly more difficult concept to embrace. Imagine the shearing faces in fig. 5.7(b) being parted as in fig. 5.7(c) then the shear strain is taken as the angle that the middle line has moved through. Unfortunately the diagram looks very unconvincing and we know from experience that the material would not distort in such a tidy way. However, if the gap is made smaller and smaller the pattern of the distortion becomes more convincing without altering the angle taken by the median line. The symbol used for shear strain is γ (read 'gamma').

5.5.6

All the above notes are simple definitions and they form the basics from which the whole analytical study of aircraft structures (the process called stressing) is built up. The next group of ideas is slightly more complicated and included for the consideration of students who may want to go on to a serious study of structures analysis.

5.5.7

In para. 5.5.1, we said that if a force or load is applied to a structural member, the member will change its length. It was discovered many years ago that if the force was doubled, the change of length was doubled. In fact the change of length is directly proportional to the change of force (up to a certain force value). This relationship is known as Hooke's Law. If we convert the force to stress and the change of length to strain, the relationship becomes even more important. Looking back at the examples in paras. 5.5.1 and 5.5.2, we worked out the stress in the member at a certain load, and the strain at the same load. If we apply half the load, we halve the stress, and according to Hooke's Law, we halve the strain. We can plot this as a simple graph as in fig. 5.10 and it becomes obvious that, arithmetically, stress divided by strain is a constant. We

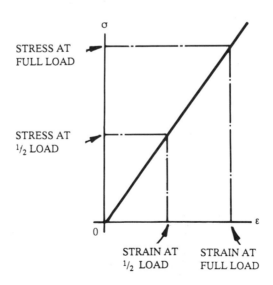

Fig. 5.10 Stress/strain relationship at small loads

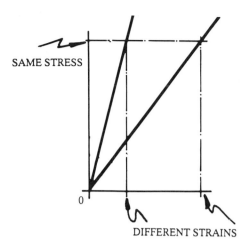

SAME STRESS

0

DIFFERENT STRAINS

Fig. 5.11   Stress/strain relationships for two different materials

said at the end of para. 5.5.2 that the stress in a member was not related to the material of the member but the same is not true of strain. Different materials extend by different amounts when subject to the same load so that relating stress and strain in the way suggested by Hooke's Law provides us with an important identifying characteristic of the material. This characteristic, or property of the material, is called *Young's Modulus* or the *Modulus of Elasticity* and using the figures from paras. 5.5.1 and 5.5.2 we can find the modulus of the material used in the examples.

1. Imperial units

   Modulus of Elasticity $(E) = \dfrac{\sigma}{\epsilon}$

   $$= \frac{27\,778 \text{ lbf/in}^2}{0.0026}$$
   $$= 10\,683\,846 \text{ lbf/in}^2$$
   $$\underline{10.7 \times 10^6 \text{ lbf/in}^2 \text{ (approx.)}}$$

2. Metric units

   Modulus of Elasticity $(E) = \dfrac{\sigma}{\epsilon}$

   $$= \frac{180 \text{ N/mm}^2}{0.0026}$$
   $$= 69\,231 \text{ N/mm}^2$$
   $$= \underline{69 \text{ kN/mm}^2 \text{ (approx.)}}$$

In chapter 6 there is a list of properties of various materials and from that we can see that the material used in our examples above is an aluminium alloy.

5.5.8

The notes in para. 5.5.7 apply to tension and compression and the Modulus of Elasticity for most aircraft structural materials is the same for both. Some materials such as fibre reinforced plastics may have different moduli relative to different directions of load application. Such materials are said to be 'not homogenous'.

5.5.9

Shear stress and shear strain have a similar relationship but instead of referring to a modulus of elasticity in shear we call this the Modulus of Rigidity or the *Shear Modulus* (symbol '*G*').

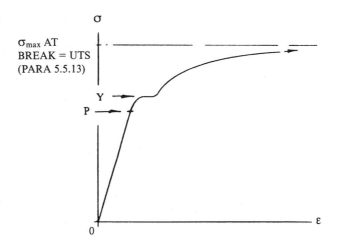

Fig. 5.12(a)   Stress/strain relationship for mild steel

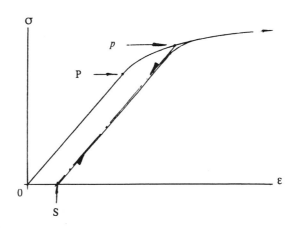

Fig. 5.12(b)   Stress/strain relationship for aluminium alloy

5.5.10

Although Hooke's Law applies to virtually all structural materials it is only true up to certain stress levels. The diagrams in figs. 5.12(a) and (b) show the type of curves obtained from tests on samples of steel (a) and aluminium alloy (b). These tests are made in a materials laboratory or test house and are typical of those made to determine whether or not a sample of material is as strong as its specification says it should be. The main curves in the diagram have been stylised and distorted a little to illustrate their characteristics, but anyone looking at actual recordings of tensile tests will be able to pick out the points which are mentioned below. The first point to notice on both diagrams is that Hooke's Law applies from O to point P called the *limit of proportionality*. After point P, as more stress is applied, the strain becomes disproportionately greater, so the curve starts to bend away from the straight line. In the case of curve (b), which is characteristic of aluminium alloy and most other structural materials apart from the softer or mild steels, the extension (strain) becomes greater and greater without much gain in applied load (stress) until the material breaks. The curve for steel is similar but has an irregularity at Y called the *yield point*, where the material suddenly stretches before starting to work properly again, which is interesting but of no great importance to the aircraft engineer simply because it only applies to mild steel. However, it gives its name, especially in American technical books, to the start of the disproportionate increase of strain.

5.5.11

In the region before the limit of proportionality (which is also called the *elastic limit*) if the material is off loaded the stress/strain curve returns to zero down the same line. The material is in fact elastic. When the stress has exceeded point P (figs. 5.12(a) and (b)) the material is said to be in a plastic state (nothing to do with plastic materials) and if the load is then taken off (say from point p in fig. 5.12(b)) the curve will not return to zero but to a new point S. At S there is no stress but there is a retained extension which is called a *permanent set*. This is an interesting and important feature for aluminium alloys because if we now reload the material we find that the elastic limit is now at a rather higher stress level. In fact the material has improved some of its characteristics slightly and has become work hardened (see also para. 6.2.2).

5.5.12

There is a further significance about point p on the curve. Structural designers in other branches of engineering usually make sure that the stress levels in their structures are kept well below the elastic limit but the aircraft designer, who is fighting a constant battle to achieve the lowest weight of structure, does not want to waste any usable load-carrying capability of the material. So, to define an upper working limit of stress value which he permits himself to use, he selects a *proof stress* (p) defined by the dimension OS when Sp is parallel to OP. When OS is 0.001 (remember ϵ is a dimensionless ratio) the stress at point p is called the 0.1% proof stress and written either as p_1 or σ_1, or when OS is 0.002 the associated stress is the 0.2% proof stress (p_2 or σ_2).

5.5.13

Before finishing this section there are some other terms connected with materials which will eventually be of interest to anyone studying structures in greater detail.

Ultimate Tensile Stress (UTS) is the stress level associated with breakage of the material and is usually quoted as a specification requirement. That is, a selected material quoted as having a UTS of (say) 54 000 lbf/in^2 or 370 N/mm^2 can be relied upon to be producing at least that stress when it breaks.

Bearing Stress is usually associated with bolted (or riveted) joints and is a value for the material being joined (not the bolt) obtained by dividing the load by the product of the diameter of the bolt hole and the thickness of the material. Permissible bearing stress or 0.1% bearing stress is quoted as a property in material specifications. Some further comment will be found in chapter 9.

In para. 5.5.1 we pointed out that when a force is applied to a piece of material of a certain length, that length changes in the direction of the applied force. For metallic materials, and most others as well, there is also a change of dimension at right angles to the direction of the force. Fig. 5.7(a) shows this obvious effect. What is less obvious is that the ratio between the strain in one direction and the strain at right angles to it is a property of the material, and is known as *Poisson's ratio* and given the symbol ν (read 'new').

Hardness of a material is measured by indenting the surface with a ball (like a ball bearing) or a point, usually a very small square pyramid, and measuring the size of the indentation. There are several systems, Brinell and Vickers being used in the United Kingdom and Rockwell being used in the USA. These hardness numbers are most useful for a quick mental comparison against known materials and there is a correlation with tensile strength (see table 6.1).

Brittleness of a material is measured by the Izod test. Of all the normal mechanical tests which are applied to materials this is the one most susceptible to small variations in the test piece. It gives an indication of 'notch' sensitivity but otherwise its usefulness is minimal.

5.6   Bending and Torsion

Now that we have listed some of the terms and expressions used we will look in more detail at the bending and torsion mentioned in paras. 5.3 and 5.4.

5.6.1

Fig. 5.13(a) shows pictorially a simple bar loaded in the middle and 5.13(b) shows the same bar diagrammatically. The most obvious point is that the bar is being bent with the top face of the bar in compression and the bottom face in tension. It is less obvious that the bar is also being sheared but consider figs. 5.13(c) and (d): it is clear that if the loaded bar was suddenly cut in either of the ways shown, there would be a movement across the cut face similar to the movement shown in fig. 5.4(b), therefore the material of the bar must be capable of resisting that movement. In the case of a plain rectangular bar there is not much doubt that the necessary resistance is there, but in some other beams, particularly those of typical aircraft construction which are built up from sheet metal and have riveted joints it is only too easy for the designer to lose sight of these shears.

5.6.2

The beam in fig. 5.14(a) is at first glance a reasonable construction with the load W being resisted by the push and pull at the lugs A and B. However, because the shear has been ignored, the vertical 'web' of the beam will collapse as shown in fig. 5.14(b). To

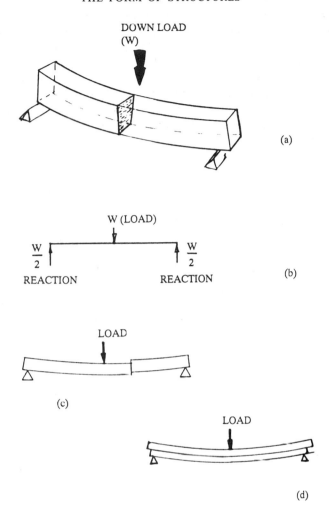

Fig. 5.13   Bending and shear due to bending

make the beam completely sound also requires an attachment at the end of the web. This would have been more obvious if the load W had been properly transferred according to the principle stated in fig. 4.2, in which case fig. 5.14(a) would have become like fig. 5.14(c). Parts (d) and (e) of the diagram illustrate alternative solutions to the problem of coping with the shear in the web.

5.6.3

Some other points should be noted about this piece of structure which as we said before is typical of aircraft structural components. It is more complex than it appears at first glance.

Firstly, the rivets through the angles attaching the web to the flanges are loaded in shear. We can imagine that if they were made of some very soft plastic instead of metal they would be cut through. Secondly, although in part (d) of fig. 5.14 we have divided the web into smaller sections, each section can still buckle into waves as shown in the

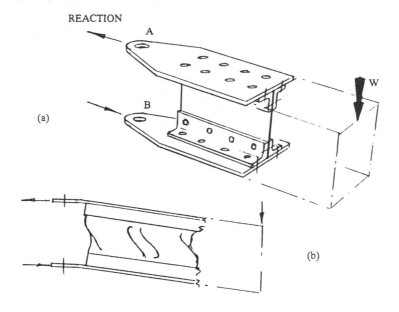

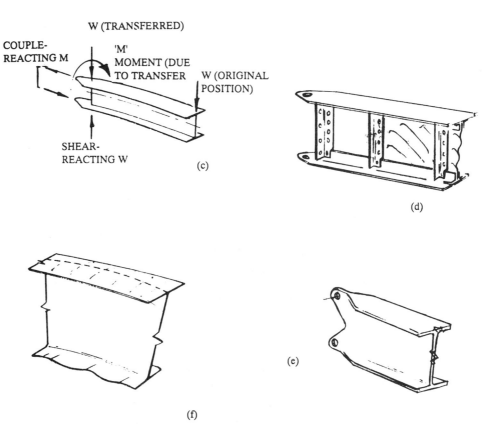

Fig. 5.14   Sheet metal beams

long unsupported web of part (b). The aircraft structures engineer in his battle against weight accepts this situation knowing that even after the thin sheet has fallen into buckles (the *post-buckling* situation) the web will still carry load. This is a very important concept which was analysed by Professor Wagner in about 1930 and followed by S. Timoshenko in 1936 whose book *The Theory of Elastic Stability*, has been the inspiration of most of the mathematical work on analysis published since.

Unfortunately, our second point above has an influence on the first point. The load on the flange rivets increases due to the action of the buckled web, and this fact gives a hint of the thoroughness with which structures of this type have to be investigated if they are to be safe and efficient.

5.6.4

Consider now the effect of making the beam deeper. Refer again to fig. 5.14(c). We will discuss the upper, tension side flange, although the stages of the argument will apply equally to the compression side, but in the opposite sense.

(a) As the beam is loaded it bends (or *deflects*) and we can see that the flange length is increased because the length round the curve of the tension side is greater than the length at the centre of the beam (which has not altered).

(b) The flange length is increased because it is loaded, therefore there is a stress (load ÷ area), and a strain (increase in length ÷ original length).

(c) If the deflection of the beam is kept the same (that is the curve of the centre is kept the same) but the depth of the beam is increased, the strain in the flange will increase and therefore, as we saw in para. 5.5.7, the stress will increase.

(d) Alternatively, if we increase the depth of the beam but keep the same strain in the flange, then the deflection of the beam will have to be less in compensation. In fact, if we double the depth of beam the deflection is half.

(e) Also if we increase the depth of the beam, the load in the flange is reduced (see fig. 4.2). Again, if we double the depth we halve the load.

(f) We know that the flange load is the area of cross-section of flange multiplied by the stress, and we have already said in (d) that we are keeping the stress constant (actually, we said we would keep the same strain but that of course means that the stress will be the same), therefore, with the same stress but a smaller load, the area and hence the weight become less.

(g) Summing up this important argument, we can say that by doubling the beam depth, we halve the deflection and halve the weight (of the flanges). It is usually said that the advantage of adding to the depth of a beam increases as the square of the depth, which is a rule of thumb which should be treated with some caution, but if the arguments above are clear and understood the effect can be considered in logical steps.

5.6.5

We will make one more point before leaving the beam in fig. 5.14. It is not a very important point because, although the type of construction is typical of aircraft structure, there are not many beams of this type which are entirely isolated; they are more usually part of a bigger structure. However, the principles involved are quite important. If the flange width is large compared with its thickness, the edges of the flange on the tension side will try to take a short cut across the curve of deflection as shown in fig. 5.14(f). This means that the strain on the edges of the flange is less than the strain at the point where the flange is attached to the web. As the strain is less we know that the stress is less and therefore the load in the whole flange is not exactly the stress multiplied by the area, or more importantly, the maximum stress is rather higher than the average found by dividing the load by the area. The phenomenon is called *shear lag*, although that term is also used in connection with the transfer of concentrated loads into aircraft skin. On the compression side of the beam, the situation is worse because the middle of the flange is being forced to reduce its length and the outer edges are unwilling to follow, so they tend to go into waves or buckles which in turn push their way towards the middle of the flange and reduce its effectiveness. In fact, if we regard the tension and compression flanges as separate members in their own right, we can say that while it is fairly easy to design a tension member, it is very difficult to design a compression member.

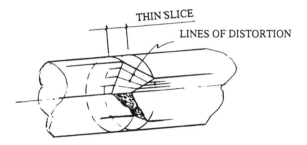

Fig. 5.15   Distortion due to twisting

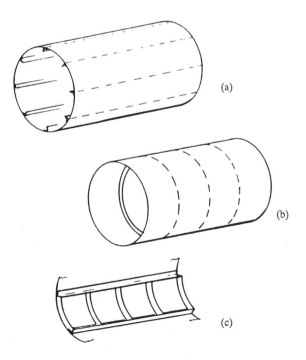

Fig. 5.16   Thin wall tubes with stiffeners

5.6.6

In fig. 5.8 we showed a solid bar being twisted or subjected to torsion. Looking at a thin slice of the bar (fig. 5.15) we can see that the metal distortion is greatest at the outside reducing to nothing at the middle. We already know that the distortion (the strain) is directly related to the stress in the material (para. 5.5.7), so that if we retain the hard working metal and dispense with the under worked metal by turning the bar into a tube we will save weight without losing much strength.

Imagine now the situation if the tube is of a large diameter and the skin is very thin. If the tube is twisted (subjected to torsion) it will carry a small load but then fall into large buckles. Those who have the strength can prove this for themselves by twisting an empty Coke can. If we now divide the surface area with relatively stiff members, as shown in figs. 5.16(a) or (b), the buckles are more contained and the load which can be carried is greatly increased. Combining figures (a) and (b) would give an illustration looking like a typical piece of aircraft structure. If we now look at just a strip of tube (which could be cut either along the tube as shown in fig. 5.16(c) or round the tube) then we have a piece of structure which is remarkably like the beam in fig. 5.14(d), and which will be just as capable of carrying loads after buckling as the beam. Before the skin starts to buckle it is carrying the load in shear. In para. 5.6.1 we saw that the material of the bar was being sheared because of bending. Conversely, in the current case, the shear due to torsion induces tension and compression 'end' loads in the stiffening members. Normally these loads balance themselves out because they act in opposite directions on either side of the stiffener member. However, they are important where the tubular member is cut out for cabin windows, access panels, aerials, etc.

5.6.7

In the paragraphs on bending and torsion (5.6.3 and 5.6.6) we made particular reference to the buckling of thin skins or beam webs and especially to their ability to carry post-buckling load. Modern aircraft generally need to maintain a smooth outer skin and it would seem strange now to see wings with rippled skins to upset the aerodynamics, but most early 'stressed skin' aircraft, many of them, like the Viscount, still flying, had wing skins which showed buckles when under load. In spite of the feeling that structures which are designed to buckle are slightly old fashioned, the concept is still important for two reasons. Firstly, in some situations, structures exploiting post-buckling strength are still designed so that the buckling commences at, or just above, proof load (see para. 4.5). Secondly, over the years since Professor Wagner's work on the subject, a great deal of mathematical analysis has been recorded on the subject of buckling and much of this has been presented in the form of *data sheets*. For the student who wishes to become a stressman, some study of the buckling of thin sheets will give him a good insight into the mathematical methods used in structures analysis. It will also give him practice in the use of the data sheets. Data concerned with buckling are as complicated as any and by practising with them he will become familiar with the use of structures data sources in general. Most major aircraft companies have their own data sheets but for people in smaller organisations the Royal Aeronautical Society and the ESDU (Engineering Sciences Data Unit) publish a very complete range.

5.6.8

In para. 5.3 we said that aircraft structures were basically boxes. If we discuss what happens to boxes when they are subjected to the types of load which they receive when they are part of an aircraft, that is, the loads we discussed in chapter 4, we shall build up an understanding of why aircraft are built in the way that they are.

The diagram fig. 5.17 is similar to fig. 4.11 and shows a representative piece of aircraft. We can call it a wing as we did in fig. 4.11 (V is lift, H is drag and T is pitching moment), or we can call it a fuselage with end ABCD built into the robust frames which carry the wings. V is now the tailplane load, H is the load on the fin and T the torque due to load H having been transferred down from its true position above the level of PQ.

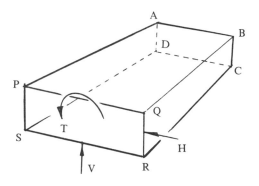

Fig. 5.17   Loads on a box structure

To help us order our thoughts we are going to treat each of the three loads (V, H and T) separately. That is, we will discuss what happens when V is applied, when H and T are each applied, and then add together the separate effects. This method of dealing with complicated load systems is called the *Principle of Superposition* and can be applied to almost any piece of structure. In general, when a structure is loaded it bends or it stretches or in some way it deflects. The load and deflection, which are there for everyone to see, go hand in hand with stress and strain which are the engineer's concern and interest. The Principle of Superposition says that if two (or more) separate loads are applied to an elastic structure, each load will produce a deflection and if the loads are applied together then the deflections will be added together. (If the deflections are in opposite directions, of course, one will be subtracted from the other.) When the loads and deflections can be added then the stress and the strain produced by the loads and deflections can also be added (or subtracted if they are in opposite directions). This principle, which is used constantly by stressmen, can go awry if any part of the structure exceeds its elastic limit, nor does it work if the deflection produced by one load upsets the ability of the structure to carry another load (see para. 11.2.2), but generally it is a useful tool which allows complex problems to be broken down into a series of simple problems.

So, looking back to fig. 5.17 and considering load V on its own, the first obvious thing to notice is that if V was applied at a point on line RS, as shown in the drawing, the thin sheet would collapse. To prevent this happening we put in a flat member

PQRS. If the structure is a wing the member would be a rib, and if the structure is a fuselage the member would be a frame; for ease in this example we will call the structure a wing. With V pushing upwards, and thinking of the whole box as a beam, face PABQ is in compression and SDCR is in tension. However, we know (para. 5.6.5 above) that the loads tend to concentrate towards the web parts of sheet metal beams, so for efficiency we should have more metal in the corners, i.e. along lines QB etc. Since in practice we have to make joints in the corners, the need for extra metal there is not inconvenient. So far then, our structure has developed into that shown in fig. 5.18. We have elected to make this example a wing, which is an advantage because if we say that plane ABCD is on the centre line of the aircraft there is a mirror image of our piece of structure the other side. This means that because the loads are symmetrical the points A, B, C and D are as rigidly held in position as they would be if they were built into a wall, and we can think of them as being so.

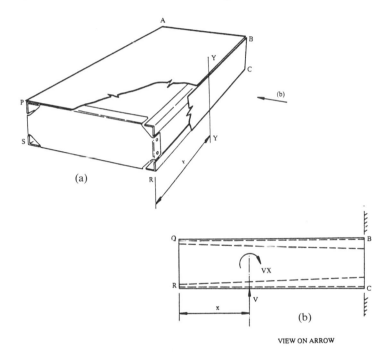

Fig. 5.18   Partially designed box structure

Remembering what was said in fig. 4.2 about transferring loads, consider the effect of transferring load V to position YY which is a distance x from QR. (Actually we should only be transferring half of V because the other half would be carried by the other side of the box.)

At YY we still have load V and we also have a moment Vx. Wherever YY is between QR and BC the load V is always the same, but the moment Vx changes from zero at QR to a maximum at BC. So if V, which is the *shear*, is constant, the web, QBCR can be the same thickness from QR to BC, but if the moment Vx, which is called the *bending moment*, increases then the material along the corners QB and BC must be more highly loaded to resist the extra moment. For efficiency we want to keep a constant high stress in the material in the corners, so the area of material has to be lower where the load is lower, and the angles shown in fig. 5.18(a) should taper as shown in fig. 5.18(b). Summarising the situation so far we have:

shear in webs ADSP and BCRQ
compression in corners PA and QB increasing towards A and B
tension in corners SD and RC increasing towards D and C.

If we now consider load *H* on its own we follow exactly the same stages of thought and conclude that we have:

shear in webs PABQ and SDCR
compression in corners PA and SD increasing towards A and D
tension in corners QB and RC increasing towards B and C.

Dealing with the torque '*T*' introduces some ideas which have not previously appeared in these notes. Earlier we discussed torsion in tubes of circular section and in para. 5.6.6 we introduced the concept of stiffeners. The present tube is just as capable of resisting or 'carrying' torsion as a circular tube. We already have jointing members in the corners QB etc., although the corner, even if it was only folded, could provide an anti-buckling stabiliser of the type which we needed to add to the circular tube. However, the large faces PABQ and SDCR will need to be divided into smaller areas to avoid large buckles or to delay the onset of buckling, since small areas of thin sheet begin to buckle later and at a higher load than large areas of the same thickness sheet. (This is an extension of the arguments in para. 5.6.3, but is fairly obvious since for the same sheet the smaller the area the greater the influence of the thickness. A 1 mm thick sheet 10 ft square would buckle more easily than a 1 mm thick sheet only 6 in square.)

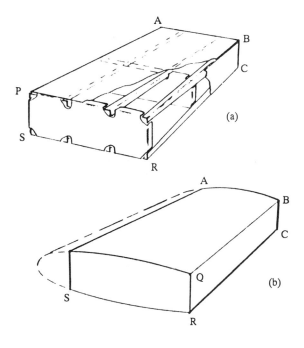

Fig. 5.19   Completed box structure

Redrawing the wing again we now have a structure as shown in fig. 5.19(a) which is now recognisable as part of an aircraft, even though in practice it would be shaped as shown in fig. 5.19(b). This diagram is in advance of the discussion because we have not finished with the torque load. An important aspect of applying torque to boxes of this shape is that they distort under load in such a way as to warp ribs such as PQRS out of plane (fig. 5.20). A rib ABCD would not warp because the other half of the wing, the other side of ABCD, produces a symmetry of loading. If there was no compensating structure or if for some reason the load on the other half wing was different (as during a rolling manoeuvre), then rib ABCD would tend to warp as well.

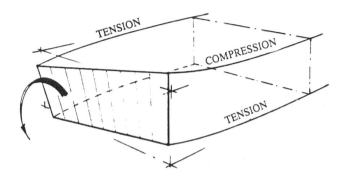

Fig. 5.20   Warping of rib or former in a thin-walled box subjected to torsion

There are at least two important considerations here. Firstly, if the rib is strong for some reason not connected with the particular case we are considering (for instance the undercarriage may be bolted to it) then it may object to being warped. In fighting against the warping loads it may produce other stresses back in the torsion box. Secondly, because it warps and is no longer a plane, flat, easy to analyse piece of structure, it may upset the principle of superposition.

Having introduced some of the side effects of torsion we can now say that the shear loads in the skin are not very different from those in the circular thin wall tube. In fact they vary according to the area of cross-section of the whole tube or box, that is, the bigger the box the lower the shear load, but the length of box makes no difference to

TABLE 5.1

	Type of end load, i.e. tension or compression			
	due to V	due to H	due to T	Type of load in total
Member PA	−	−	0	Large compressive load
QB	−	+	0	Smaller load
RC	+	+	0	Large tensile load
SD	+	−	0	Smaller load
		Type of shear load		
Skin PQBA	0	+	+	High shear load
QRCB	+	0	+	High shear load
SRCD	0	−	+	Lower shear load
SPAD	−	0	+	Lower shear load

the shear load. (Note, however it will make a difference to the deflection and the warping.) The shear due to torsion acts all round the box but the shear due to V and the shear due to H each only act on two faces. Looking back to fig. 5.17 we can see that the shear due to T increases the shear in PQBA which was due to H but reduces the shear in SRCD, and there is a similar effect in the shears due to V. We can now add together all the effects of V, H and T for the case we have considered and the results are shown in table 5.1. The shear due to load T we will call $+$ (read 'positive') for this example. Tension in a member we will call $+$ and compression will be $-$ (read 'negative'). Without putting physical dimensions into the example we cannot say whether the members (also called edge members or booms) QB and SD are in tension or compression. This depends on the relative sizes of the loads due to V and H.

5.6.9

Before proceeding further we must sound some warnings. The example above is not 'stressing' which is more mathematical and much more searching. Nor is it really representative because a wing would not, for instance, have all its lift (load V) concentrated at a single point. However, as an example it introduces some of the concepts associated with box structures in torsion and bending and gives a hint of the way a stressman might begin to approach the problems of analysis. His later stages would involve detailed examination of the skins and members to see if each of these could carry the loads imposed on it; examination of the loads in the rivets or other fasteners in the joints; reallocation of loads if he has been obliged to strengthen one part and alter the strain on another. After all this, remember, he has only examined the structure under one loading case with all other cases still to come.

5.6.10

In an effort to present an uncluttered broad picture many important details have been glossed over. We will now discuss one which is very important. In fig. 5.17 we had the point of application of the loads centrally placed and it was easy to accept that the influence of the loads was not biased towards one or other piece of structure. Three situations could have upset this state of affairs: either the structure could have been asymmetrical, or the point of application of the load could have been elsewhere or we could have had both of these situations simultaneously.

If the point of application of the load is not convenient we can shift it to any other position we like provided that we add a moment (see fig. 4.2). In fact in fig. 5.17 we had a moment 'T' which could have resulted from a movement of V.

The question of asymmetry is more complex, and we are now expanding the discussion away from box structures and on to beams of any sort. We can see that if the structure in fig. 5.17 had been stiffer on the PS side than the QR side (say that members PA and SD were a heavier section than members QB and RC), application of load V at the point shown would have resulted in point Q deflecting upwards more than point P. That is, the rib PQRS would have rotated anticlockwise as well as deflecting upwards. In fact we are saying that load V would have twisted the box, or applied a torsion to the box, as well as applying a bending. To benefit from using the Principle of Superposition for the analysis of beams we need to find a position to which we can move V, so that it bends the beam without twisting it. (Remember we can move V where we like by adding a moment which will change T.) Such a position is called the *flexural centre* or *shear centre* and is a property of the cross-section of beam under

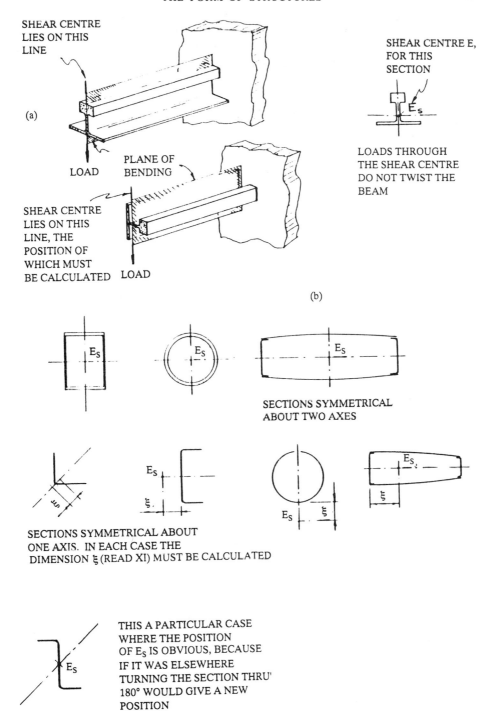

SHEAR CENTRE
LIES ON THIS
LINE

(a)

LOAD

PLANE OF
BENDING

SHEAR CENTRE
LIES ON THIS
LINE, THE
POSITION OF
WHICH MUST
BE CALCULATED   LOAD

(b)

SHEAR CENTRE E,
FOR THIS
SECTION

E_s

LOADS THROUGH
THE SHEAR CENTRE
DO NOT TWIST THE
BEAM

E_s E_s E_s

SECTIONS SYMMETRICAL
ABOUT TWO AXES

E_s E_s E_s
ξ E_s ξ ξ

SECTIONS SYMMETRICAL ABOUT
ONE AXIS. IN EACH CASE THE
DIMENSION ξ (READ XI) MUST BE CALCULATED

E_s

THIS A PARTICULAR CASE
WHERE THE POSITION
OF E_s IS OBVIOUS, BECAUSE
IF IT WAS ELSEWHERE
TURNING THE SECTION THRU'
180° WOULD GIVE A NEW
POSITION

E_s $= $ E_s

Fig. 5.21   Shear centre and its position

consideration. When the section of the beam is symmetrical about the plane of bending, the shear centre is on the line of symmetry. When the section is not symmetrical the position of the shear centre has to be calculated.

In fig. 5.21(a) we have arranged the plane of bending to pass through the shear centre. If the bending load had been applied elsewhere we would have shifted it but of course the particular section shown is not a good shape to resist the torsion which such a shift would have produced.

Fig. 5.21 shows some sections with a rough indication of their shear centre. The position of the shear centre of the channel section is of some interest. Because it is easily made and easily attached the channel section is a popular structural member particularly for mounting pieces of equipment and for other jobs which at first glance appear to be quite simple. Unfortunately loading the member in what appears to be a straightforward manner, such as that shown in fig. 5.22, produces a twisting action. A situation of this sort would not arise with a box section or a tube because the shear centre would be within the section and such sections are good torsion carriers anyway. Remember, the channel is a very poor torsion carrier and should only be used with caution.

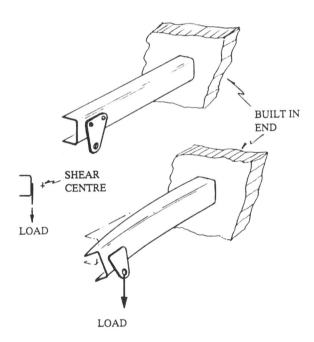

Fig. 5.22   Effect of load on a channel section beam

5.6.11

So far all of this section has been concerned with boxes made of thin sheet metal and of course this is the most familiar type of structure used on modern aircraft. Variants are also in general use where the webs and booms (see fig. 9.12) are machined out of very large thick plates of metal. This method goes a long way to avoiding potential breakage (failure) points where rivet holes are drilled or other discontinuities of section occur in a built up structure. More comment on this will be found in chapter 9. However, this type of structure is basically the same as the stressed skin structure, but historically there were other structures which occasionally reappear in new designs. Looking back at fig. 5.19 the element could have looked like part of fig. 2.2 or if it had been part of a fuselage instead of a wing it could have been like fig. 5.2. One of the most interesting boxes, shown in fig. 2.1 was used on the first successful flyer and subsequently on many other aircraft.

5.6.12

When looking at a piece of aircraft structure, however important or insignificant, remember that it almost certainly will have to carry bending and torsion, so it should be, in some form, a box or tube. If it is not a box and is perhaps only carrying some local load then look for some other method of stabilisation, especially if it is a channel section or an angle section.

5.7 Compression

5.7.1

In para. 5.6.5 we mentioned compression load effects in the flanges of beams. To have made the example structure in para. 5.6.8 more representative of a general case we could have added a compressive (or end load) to the box, but the larger elements of aircraft structure do not usually suffer much from loads along their main axis. Fuselages are in tension from cabin pressurisation loads and, for some aircraft, in compression from engine loads, but usually end loads are a problem for parts of the boxes rather than the whole. In our example we had boom PA suffering compression which would produce a buckling problem. In the present case the buckling and stability problem is slightly different. Previously we discussed buckling in panels due to shear and in para. 5.6.5 buckling in compression flanges of beams. In both of those cases the buckles were contained and forced to fade out by stiff edge members or by the web in a beam, so that under load the structure would sense the sudden formation of a buckle and 'give' a little, but then the load would be redistributed into other paths. Buckling of compression members is usually more final. In the case of the boom PA, for instance, if it chose to buckle (or collapse, or cripple) inwards, that would be final.

(The classic mathematical analysis of compression members or struts was the work of Euler (read oiler) in about 1750. The basis of his argument is that a slim straight springy strut (the usual modern given example is a bicycle wheel spoke) will support a compressive end load even though the strut is slightly bowed. Many students find difficulty in accepting this basic premise because practical experience suggests that if they press down on a spoke it will support a small load and then suddenly buckle and give way. If we think of it another way and start with a small sideways bow in the strut while pressing lightly down on the top we will feel the strut pushing back as it tries to straighten up. The down load on the strut is the Euler buckling load. Under this load the strut is in equilibrium and any increase of load causes a crippling failure.)

The strength versus weight problems associated with the design of struts were a major worry in early aircraft structures. In a stressed skin type of structure if a part is in compression and the calculations suggest that it may be in danger of buckling under an extreme loading case, it is usually possible to add some stabilising auxiliary structure to avoid the problem. In the example shown in fig. 5.17 if PA looked as though it might fail we would probably put an extra rib, of lighter construction than rib PQRS, half way along the box. This would stabilise PA considerably. Solutions of this type were not open to designers of biplane interplane struts. However, stressmen and designers do still come up against strut problems on modern aircraft in the mounting of equipment and in non-stressed skin areas such as undercarriages.

5.7.2

Another area where there is a compression/buckling/stability problem is in riveted joints at the edge of shear carrying panels. As we noted in para. 5.6.3 the rivets attaching beam flanges to beam webs are loaded in shear and the same applies to all the rivets around the edge of a skin panel. If each rivet is loaded then there must be end loads in each piece of metal between the rivets. This end load can produce *inter-rivet buckling* in the skin, an undesirable thing in itself but worse because the deformed skin can pry (i.e. force up by leverage) under the rivet heads which are not designed to carry that sort of load.

5.8   The Whole Structure

We have discussed the general form of aircraft structure and the way it works and to complete this section we should record a few definitions of phrases used in connection with structure design.

5.8.1

Primary structure is structure which, if it failed, would endanger the aircraft.

5.8.2

Class 1 parts are those components, the failure of which could cause structure collapse, loss of aircraft control, injury to occupants or interference with essential systems.

5.8.3

Class 2 parts are those not included in Class 1 which require more than visual examination to ensure reliability.

5.8.4

Class 3 parts are the remainder.

5.8.5

Fail-safe structure is so designed that after failure of one part there is still sufficient strength in the remainder to prevent catastrophic collapse until the damaged part is found by routine inspection.

5.8.6

The above definitions were paraphrased from BCAR which are now replaced by JAR (see p. 00) and the current description for an important piece of structure is 'Critical Part'. However, the earlier phrases create a clear picture in the mind for aircraft structural engineers who are really only concerned with primary structure and Class 1 parts. Fail-safe concepts are usually linked with fatigue failures and modern thinking favours easier inspection of components likely to suffer from fatigue damage, thereby reducing the need for alternative load paths. The whole problem of evaluating the degree of success in a 'fail-safe' design is very difficult and well beyond the scope of this book.

5.9 References

Timoshenko, S., *Theory of Elastic Stability*, New York: McGraw-Hill Book Co.

(See also books listed in para. 11.5.)

CHAPTER 6

Materials

6.1 Choice of Materials

6.1.1

With the exception of reinforced concrete, in aircraft construction you can find almost all the engineering materials used by man. The original aeroplanes were made mainly of wood with fabric covering and steel wire bracing; today the main material used is *aluminium alloy*, which is pure aluminium mixed with other metals to improve its strength. In other parts of the aircraft, for instance in the avionics, valuable metals such as gold and platinum can be found; plastics are used in the furnishings, both in sheet form and as upholstery foam, and in recent years even paper has been employed, for instance as the centre of sandwich structures used in the floor of wide-bodied aeroplanes.

With so many choices open to him the aircraft engineer needs to be very selective when considering which material to use for the job that he has in hand. Selection of some materials is very obvious and scarcely needs recording. For instance the gases coming from a jet engine are extremely hot and therefore it is not possible to use any material that is likely to be damaged by heat in any position where it can be touched by jet efflux. It is also ridiculous to suggest that anything other than a transparent material could be used for windscreens or cabin windows, but nevertheless it is not particularly easy to find a transparent material which will stand the loads and extremes of temperature to which such components are subjected. At other times, choice of materials may be determined or restricted by the manufacturing processes likely to be employed on them. For instance, not all steels are suitable for welding. As a process, welding is not much used in aircraft construction now, but in cases where it is, the particular alloys used must be selected with considerable care. In some cases aluminium alloys have to be softened before they can be bent or formed into shape. This softening process may be accompanied by a later heat treatment process to regain the former strength of the alloy. If this is the case it is often impossible to perform the heat treatment after the material has been bent or formed due to the effects of distortion, and it will then be necessary to choose another alloy. Amongst the different types of materials available to the engineer, there are an enormous number of sub-variants. The properties and characteristics of these subvariants are carefully controlled, and the materials manufacturers produce their products to written specification which may be issued by the standards institutions of the various countries in which they are made, or by the aircraft industry itself.

6.1.2

By far the most common material used in aeroplane structures is aluminium alloy, also called *light alloy*. A typical light alloy used in aircraft structures may be 4% copper, $\frac{1}{2}$% silicon, $\frac{1}{2}$% magnesium, a little manganese, iron and chromium and the remainder, more than 90%, will be aluminium itself. Mixing the aluminium with these very small quantities of other metals has an enormous effect on its strength; the strength of the structural alloys is six to eight times the strength of pure aluminium. When it was first discovered, aluminium was very difficult to refine and for some years in the middle of the 19th century it was virtually a semi-precious metal. Even when production methods were improved and cheapened it was of very little use as a structural material because of its weakness; it was not until about the time of the First World War that the discovery of methods of producing alloys presented the opportunity for the design of the type of structures which have changed very little up to the present time. By far the most important design objective for the aircraft structural engineer is lightness, but of course it must be coupled with adequate strength. We discussed in the last chapter the meaning of the word strength. Strangely enough, in terms of tensile strength, there is very little benefit to be gained from using light alloy. Although it is only one third of the weight of steel, it also has only one third of the strength. However, for other types of structural members, such as bending members or struts, it is useful to have a bulk of material and it is then that aluminium alloy comes into its own. This need for bulk is very well illustrated by the skin sheeting used on fuselages and wings. The thickness of the skin on quite a large aeroplane may be as little as 0.030 to 0.040 of an inch (1 mm). Much of the same strength of skin could be achieved with steel a third of the thickness without any loss of weight, but clearly skins which are only 0.010 of an inch (0.3 mm) thick would not be a very practical proposition.

Aluminium alloy starts life as a cast billet. There are many uses for cast aluminium but in general it is not a good structural material. Cast aluminium is widely used on reciprocating engines (crank cases and pistons) but it is not a ductile material. This lack of ductility, which is another way of saying it is rather brittle, makes it unsuitable for many structural applications and this unsuitability is recognised by the Airworthiness Authorities (see chapter 10) who ask for considerable reserves of strength to be applied to any structural members using castings. More usually, the cast billet is worked on and submitted to various processes which change its character into what is termed a *wrought alloy*. Wrought alloys are used in several forms:

    sheet and plate
    forgings
    extrusions
    tubes

In the process of changing from cast to wrought material the cast billet, which is probably a cylinder perhaps 10 in diameter and 25 in long, is first subjected to a repeated squeezing process not unlike kneading bread. The object of this is to change the random crumbly structure of the casting into a material with a grain or a recognisably continuous 'bound together' type of internal texture. To make sheet or plate, the kneaded block or billet is passed through a succession of heavy rollers which reduce the thickness of the material to usable and carefully controlled dimensions. For forging, the large billet is cut into smaller pieces which are then, quite literally, hammered and squeezed into shape, in most cases between carefully prepared dies, to produce the finished part. Components which have been forged are sometimes difficult to distinguish externally from components which have been cast, but the manufacturing processes of forging produces a much better quality of material than the casting process. The process of making extrusions is often referred to as being like squeezing toothpaste out of a tube and this is a good analogy, with the nozzle of the toothpaste tube being shaped to produce long strips of material of complex sections. It may be difficult to believe that this is literally what happens with a metal, but given that the machines are very robust, this is indeed the process. Tubes start life as hollow extrusions which are subsequently drawn through circular dies. This drawing process again improves the quality of the material as well as thinning it. Accepting that light alloy is the main material used in the construction of aeroplane structures, we can list its major advantages as follows:

It is a light material especially when used in bending or compression members which form the major part of any aircraft structure.

It is easily produced in usable forms.

It is also a material which is comparatively easy to work. (Chapter 7 discusses some of the manufacturing processes used.)

However, materials which have advantages also have disadvantages, and light alloy is no exception. One of the major problems is a process known as ageing. Further on in this chapter there are some notes on the heat treatment of aluminium alloys, and it will be seen that part of the process of improving the strength of copper bearing alloys is a process known as *precipitation* or artificial ageing. Unfortunately this process continues throughout the life of the material and although modern materials are less susceptible to this changing than older alloys were, a material carefully selected today may be found to have different properties in ten years' time. Another disadvantage with aluminium alloy which has become well publicised in recent years is the phenomenon known as *fatigue*. Fatigue occurs eventually in metal members which are subjected to varying stresses, especially where the stress varies from tension to compression in a cyclic way. A large number of cycles would be involved before failure. Unfortunately the nature of aircraft structures being as it is, with constant pressurisation and de-pressurisation of cabins, bending stresses produced by gusts on wings, and perhaps worst of all, stresses produced by vibration in the region of engines or helicopter blades, fatigue becomes a major problem for the aircraft structure designer. Basically, the problem can be dealt with by limiting the stress on a member to a figure which is considerably less than the maximum stress permittd by the specification of the material. This fatigue limit stress lies in the region of 10 ton/in² for almost any light alloy, and designers using wrought alloys at higher stress levels have to consider very carefully the number of fatigue cycles to which the member is likely to be subjected. A great deal of research has been put into the phenomenon of fatigue, and published literature is widely available to the serious student. Possibly an unfair impression has been given that aluminium alloys are more susceptible to fatigue than other metals. This is not necessarily so but aircraft design is always so much closer to the limits of technology, research and performance, that aluminium alloys are always in the forefront when fatigue is being discussed. A third disadvantage with aluminium is its susceptibility to the effects of notching, or small cut-outs, and this is discussed in more detail in chapter 9.

6.2   Material Properties

In specialist books dealing with metallurgy, it is usual to list metals in a certain order which is followed here.

6.2.1

Steel is iron with other elements added to make an alloy. The alloying elements are present in a vast variety of different combinations and quantities to make a range of materials fitted for many different tasks. The scope is so wide and we are so familiar with the material that is requires some mental effort to realise that the rusty reinforcing bars sunk into the concrete foundations of a building are made of the same basic material as the surgeon's scalpel. When the aircraft industry developed beyond its initial problem of achieving sustained flight it quickly became aware that wooden constructions deteriorated due to moisture absorption. The obvious step was to turn to the universal material of engineering construction and the earliest metal aircraft used steel members where wood had been used before, but retained the fabric covering. Fig. 2.3 shows a structure of this type with a tubular steel framework quite clearly similar to the wooden frame of the earliest flying machines. (The first all-metal aeroplane was the Tubavion, a French design of 1912, but the more usually accepted first successful 'all-metal' aircraft was the J1 built by Professor Junkers in 1915.)

Subsequently, the amount of steel used in aircraft structures has been steadily reduced to the point where the designer uses it only with reluctance when he cannot find another material to suit his purpose. This is partly due to improvements in the properties of aluminium alloys and the use of titanium but more importantly because structures made of mixed materials are generally best avoided. With aircraft at their present size it is not sensible to consider an aircraft with an all-steel outer covering (see para. 6.1.2) so if the skin is aluminium alloy, that tends to dictate the material of the whole structure. There are two reasons for avoiding mixed materials: one is a corrosion problem which is mentioned in chapter 8, but a more important reason is that steel is stiffer than aluminium (higher Young's Modulus) and so, as will be mentioned in para 9.9.2, it will try to carry all the load, leaving the aluminium underworked, and therefore inefficient. Nevertheless, steel is still an important aircraft material and is used in undercarriages, engine mountings, joint plates (at wing roots, etc.), door latches, for bolts and fasteners, and in numerous other places where its valuable, well-documented properties, its reliability and its comparative cheapness outweigh its disadvantages of high weight, corrosive tendencies and high fabrication costs.

(The 'high weight' comment above requires some qualification. In general, steel is three times the weight of aluminium but is also three times as strong so there is no intrinsic weight penalty in the use of steel provided that all the material used is fully worked. Unfortunately, this ideal situation is virtually impossible to achieve. For example, the head of a steel bolt carries material that is only there for the benefit of the spanner which tightens the bolt perhaps only once in its life. It is really that sort of waste and even larger volumes of low stressed material in more complex fittings which account for the 'high weight' of steel.)

The principal alloying element added to iron to make steel is carbon. This addition of carbon to pure iron is virtually a definition of steel, although in recent years some strong steels have been developed which have only very small quantities of carbon (see 'maraging steels' below). In general, the greater the carbon content the stronger the steel. The student reading text-books on metallurgy will be confused to find that cast iron, which is a relatively weak material, contains between 2.5% and 3.5% of carbon, while the carbon content of a low strength mild steel is about 0.2%, and in the case of a high strength alloy steel it is about 0.5%. The situation is that in the two materials the carbon is in different forms. In cast iron it exists as graphite suspended in the pure iron like particles in muddy water, while in the steel it is truly in solution with the iron. The other major alloying elements are:

manganese (used between $\frac{1}{2}$% and 2%)
chromium (may be up to 25% in some heat-resistant stainless or corrosion-resistant
    steels)
nickel (can be as high as 20%)
molybdenum (0.1% to 0.8%)
silicon
vanadium
sulphur
phosphorus

There is a vast range of steel specifications (internationally there are several hundred) each having properties making it suitable for a certain task. These various specifications are very loosely grouped into:

carbon steels – low carbon
                      – high carbon
alloy steels
stainless or corrosion-resistant steels
heat-resistant steels and super alloys

The stainless and heat-resistant steels speak for themselves and are not of much interest now to the structures departments of the aircraft industry. Although a few years ago the Bristol Aeroplane Company built an all stainless steel fighter size supersonic research aircraft called the Type 188, a similar project now would probably use titanium. Stainless steels may be specified for some new projects where its comparatively low cost relative to titanium may still be attractive but this is becoming less likely (see para. 6.3 below).

Low-carbon steels are the group which go to make bridges and horseshoes and almost every other engineering construction in the world but they are not of use to the aircraft engineer. Although they have a much higher Young's Modulus than light alloy (30×10^6 lbf/in^2 as against 10.5×10^6 lbf/in^2) their ultimate tensile stress, about 21 ton/in^2 or 325 N/mm^2, compares unfavourably with fairly uncomplicated light alloys at about 25 ton/in^2 (385 N/mm^2).

The remaining carbon and alloy steels are used in aircraft structures mainly for fittings and fasteners. The usual group names are very vague and really only jargon; for the aircraft engineer it is more convenient to group the specifications into those which are weldable (or not weldable), those which are difficult to machine, etc., and to grade them all in order of strength. Nevertheless, we should attempt a definition and say that alloy steels can be considered as those with total alloying constituents above 5%, and high-carbon steels those with carbon above 0.2% and with manganese as the only other constituent. It will be seen from table 6.1 that the range of strengths available is wide and that even when requiring a property as particular as a welding suitability, the designer is still faced with a choice and therefore the need to make a decision. A total list of available steel specifications would increase the problem so that the short list shown should only be taken as a general indication of the values of mechanical properties. A designer or stressman would know (or would find out) which steels were in use or in stock within his company.

In recent years a new range of steels has been evolved called *maraging steels* from which carbon has been virtually eliminated. The alloy steels which achieve their strength with a high carbon content are usually very difficult to machine (note, the Brinell hardness number, in table 6.1, for S.28 and S.98) and often rather brittle or notch sensitive (note the Izod number in table 6.1). The maraging steels are strong, tough and comparatively easy to work. Their heat treatment process is different from carbon steel and simpler (see para. 6.4 below) but their initial cost is high. This may be offset against lower machining costs but the total cost of any particular component in maraging steel should be compared with the cost in titanium, also bearing in mind the considerations of para. 6.3 below.

6.2.2

Non-ferrous metals are usually taken to mean copper and the copper bearing alloys: brass, bronze and gunmetal. None of these are of specific interest to the aircraft structures engineer, but copper has an interesting physical property known to metal workers for many years which also exists in aluminium and is exploited in some aircraft manufacturing processes. Copper that is stretched or is deformed by hammering becomes tougher in a process known as *work hardening*. The phenomenon is only mentioned here to introduce the term; its usefulness in working aluminium will be discussed in chapter 7.

6.2.3

Aluminium and its alloys have been mentioned in para. 6.1 above on Choice of Materials. It seems certain that these will remain as the major materials of aircraft construction for many years to come, although, in percentage terms, titanium and reinforced plastics will make progressively larger inroads. There is also, at present, a limitation on the use of light alloy on supersonic aircraft because at high speed the skin of the aircraft can approach temperatures which are detrimental to the strength of the material. Concorde was designed as a Mach 2.2 aeroplane because higher speeds would have dictated the use of a material other than aluminium alloy. Even then a special alloy was developed particularly for the aircraft with the object of obtaining an improvement in the strength and fatigue characteristics over a wide range of temperatures.

In the notes on steel in para. 6.2.1 we said that steel was graded or denominated by its carbon content, and then by the quantity of other elements in its composition. A similar system applies to light alloy, which falls into three groups. The original formulation was patented and marketed under the trade name of Duralumin. For many years the name 'Dural' was freely applied to all varieties of light alloy (at least in

TABLE 6.1 Steels

Country of origin	Ref. no.	Permissible tensile stress[2]		Elongation at break (%)	Hardness (para. 5.5.13) Brinell (B) or Rockwell (R) B or C	Izod. (para. 5.5.13)	Remarks
		at yield (para. 5.5.12)	ultimate (UTS) (para. 5.5.13)				
UK	BS.S.1	51–65 000	78 000	10	(B) 160	15	0.20% carbon lowest grade steel for aircraft use
UK	S.92 / S.514 / T.45	69 000 / 90 000 / 90 000	90 000 / 112 000 / 100 000	15 / 10	(B) 180–250 / (B) 230	35 / — / —	Bar / Sheet / Tube } Suitable for welding and compatible
UK	S.154	96 000	127–156 000	12	(B) 250–300	40	General purpose bar
UK	S.98	145 000	168 000	10	(B) 360	25	High strength bar
UK	S.28	152 000	224 000	8	(B) 445	15	V. high strength alloy bar, but note Izod trend
UK	S.80 / S.521	98 000 / 29 000	127 000 / 78 000	12 / 30	(B) 288 / —	25 / —	Bar / Sheet } General purpose CRES. Sheet is not structural but good forming properties
USA	SAE.1015	45 000	55 000		(R) B64		0.15% carbon lowest grade suitable for aircraft
USA	4130 / 4130	70 000 / 103 000	90 000 / 125 000	17 / 10	(R) B90 / (R) C26		General purpose steel in various forms and HT conditions. The major a/c steel
USA	4340	145 000	180 000	14	(R) C40		Compare with BS. S.98

NOTES TO TABLE 6.1

1. The properties shown are only approximate and for design purposes reference must be made to the published specification.
2. Stress is quoted in lbf/in². To convert to N/mm² divide the figure given by 145.0 (see also note 5).
3. Young's Modulus for steel is 28–30 × 10⁶ lbf/in² (read twenty eight to thirty by ten to the sixth pounds force per square inch).
4. Poisson's ratio for steel is 0.27.
5. Stress may be quoted in units called kips, 1 kilo pound = 1000 lb.

TABLE 6.2 Light alloys

Country of origin	Ref. no.	Permissible tensile stress		Elongation at break (%)	Brinell hardness no. (para. 5.5.13)	Remarks
		0.2% proof (para. 5.5.12)	ultimate (UTS) (para. 5.5.13)			
USA UK UK	2014–T4 L.164 L.102	35 000	56 000	14	105	L.164 is sheet material L.102 is a bar
USA UK	2014–T6 L.165	50 000	61 000	8	135	Note the effect of heat treatment on properties. UTS increases 9%, proof stress increases 43%
USA UK	2024–T4 L.109	39 000	59 000	15	120	
USA UK UK UK	QQ–A–250/11 L.111 L.113 L.114	37 000	43 000	8		L.111 is bar } This is weldable alloy. L.113 is sheet } Properties are before L.114 is tube } welding
USA UK	7075–T6 L.88	62 000	71 000	8	150	Aluminium – zinc alloy (see para. 6.2.3)

NOTES: 1. The properties are approximated to illustrate the grouping. Figures must not be used for design and reference must be made to the specifications.
2. Stress is quoted in lbf/in². To convert to N/mm² divide the figure given by 145.0 (see also note 5).
3. Young's Modulus for light alloy is 9.5 – 10.7 × 10⁶ lbf/in².
4. Poisson's ratio for light alloy is 0.3.
5. Stress may be quoted in units called 'kips'. 1 kilo pound = 1000 lb.

Europe) in much the same way that the name 'Hoover' was applied to all makes of vacuum cleaner. Dural is a copper bearing alloy which was reputedly discovered by accident and possessed the rather strange characteristic that if it was heated (to about 480° C) and suddenly quenched in water it became soft and easily bent or formed, but then the formed shapes, over a period of a few hours, regained their former strength and hardness naturally and without any further attention. This remarkable property, which is entirely different from the work hardening characteristic of copper and brass (although some aluminium alloys also work harden), has influenced the techniques, and hence the design philosophy, of aircraft manufacture for sixty years. It was later found that a further heat treatment, which was called artificial ageing (at about 170° C), made an additional improvement in the strength. For comparison the 0.1% proof stress of naturally aged Dural was about 33 000 lbf/in^2.(370 N/mm^2). Modern derivatives still have much the same strength.

The artificially aged copper bearing alloy was well established by the early 1930s, with the second group of alloys already making an appearance. This group differed from the Duralumin group by the addition of nickel and a higher magnesium content. Not widely used for structures, this group was mainly employed in aero engines and as forgings.

The third group, which came into use in the middle 1940s, used zinc and magnesium as alloying elements. The 0.1% proof stress of a typical early member of this group was 74 000 lbf/in^2 (510 N/mm^2) which represented an enormous increase over the previous families of alloys. Unfortunately, possibly because wartime pressures had influenced the development programme, the desirable properties were accompanied by some highly undesirable tendencies. Fatigue cracking (see para. 6.1.2) accompanied by high notch sensitivity (para. 9.3) led to some unpleasant failures. A new phenomenon called *stress corrosion*, initiated by unrelieved internal stresses caused by heat treatment, led to cracking which was very difficult to detect. Many designers considered that the disadvantages outweighed the benefits, but recently improved versions have brought this group of alloys into more general use.

Table 6.2 lists the mechanical properties of some of the widely used alloys. The designation used is nominally international but most familiar in America. The current equivalent British specifications can easily be checked in the brochures published by the materials producers. Some notes on the various methods of numbering specifications are given in para. 6.5.

6.2.4

Titanium is effectively the last on the list of metals used in aircraft structures, although magnesium was used in the past and beryllium may be used in the future. One of the most plentiful metals on earth, it is difficult to refine and therefore expensive. Its weight is between that of steel and light alloy, but it is strong and corrosion resistant even at high temperatures. These characteristics have led to its increasing use for specialised pieces of structure, such as fairings near jet engine efflux or intake edges where stainless steel may have been used previously. On the Boeing 727 designed in the early 1960s, less than 2% of the structure weight was titanium, but by the late 1960s the Boeing 747 structure was almost 10% titanium by weight. The trend has been matched by other civil aircraft and almost certainly exceeded by military designs. It is interesting to speculate how much the designers of the Boeing 747 were driven, against their will, to use titanium because weight in the final design exceeded that of the project calculations. If they were forced to use more titanium than they would have wished, then the price of the aircraft must have been higher than expected; but in the event it is a magnificent money making machine and supports the argument that the cost of raw material should not necessarily be a deterrent to its use. (See also para. 6.3.) As with other metals titanium is used as an alloy, in fact as two families of alloys, one with aluminium and one with manganese. These give a good range of available properties (see table 6.3) which should be compared with those in tables 6.1 and 6.2.

TABLE 6.3   Titanium and maraging steel

Country of origin	Typical ref. no.	Description	Permissible tensile stress		Elongation at break (%)
			0.2% proof (para. 5.5.12)	ultimate (UTS) (para. 5.5.13)	
UK	T.A.10	Titanium sheet	130 000	140 000	8
UK	T.A.6	Titanium sheet	67 200	82 900	15
US	AMS.4935	Titanium sheet	130 000	140 000	8
UK	DTD.5212	Bar ⎱ maraging ⎰ steel	245 000	260 000–290 000	8
US	AMS.6521	Sheet ⎰	270 000	280 000	8

NOTES:
1. The properties shown are approximated to illustrate the comparison with the materials shown in tables 6.1 and 6.2. Figures must not be used for design and reference must be made to the specifications.
2. Stress is quoted in lbf/in^2. To convert to N/mm^2 divide the figure given by 145.0 (see also note 4).
3. Young's Modulus for maraging steel is 27.5×10^6 lbf/in^2. Young's Modulus for titanium is $15.5 - 16 \times 10^6$ lbf/in^2.
4. In some lists of material properties stress is quoted in the units called 'kips'. 1 kilo pound = 1000 lb.

6.2.5

Reinforced plastics and bonded structures are now widely used in aircraft structures especially in floors and ancillary structures such as overhead luggage bins.

The most familiar of these materials is *Fibreglass* which is another of those names like Dural and Hoover where a trade name has become a part of the language. More correctly Fibreglass is one company's name for a loose mat of random fibres of glass. The glass is spun, in the sense that spiders spin the threads of their webs, by passing molten glass through very small holes. These very thin fibres are also spun in the other sense, like cotton or wool fibres, into threads (or *rovings*) which can be woven into cloth. Because the glass is spun so fine, the threads can be bent to a radius of the same order as the thickness of the cloth so that a brittle material has been changed into a flexible material without losing its strength or resistance to corrosion.

What is usually known as fibreglass is, in fact, *glass reinforced plastic* (GRP), manufactured by laying glass cloth, or glass random mat, onto a prepared surface and then impregnating the cloth with liquid plastic which sets hard into a stiff composite sheet.

Two groups of plastics are in common use for making GRP. *Polyester synthetic resin* is the more common and *epoxy resin* is the more expensive but has rather superior properties. The highest strength glass reinforced layups use a special cloth called *unidirectional* which is not woven but laid as carefully prepared papers of lightly bonded threads. Each layer of paper has its threads running parallel to one another and successive layers are placed at right angles to each other like the grain in plywood. The resulting sheet is often used as the facing for a *sandwich panel* board, the interior of the board being a honeycomb of plastic impregnated paper (see also para. 5.2). Used in this way reinforced plastics are very efficient in terms of strength and lightness, and components grouped under the general heading of composite and bonded structures are becoming more and more accepted. Helicopters have been built using GRP in their primary structures including rotor blades, but fixed wing aircraft designers seem to reserve the reinforced plastics for floors, radomes and non-structural ducting, etc., although the military use may be greater. The Northrop company said that by the mid-eighties 55% of their structures would be composite.

(It is not the intention to give the impression that all bonded structures are necessarily either plastics or sandwich structures. Metal faced, metal honeycomb cored panels and metal panels with adhesives are all bonded structures and in the early days of flying, wooden laminated structures bonded with casein glue were used for some excellent aircraft, among them the beautiful Lockheed Vega. A later version using plywood skins on balsa wood core bonded with synthetic resin glue was used for the structure of the even more elegant De Havilland Mosquito.)

6.2.6

Plastic is also used without reinforcement but other than for windows it is not normally used as part of the structure.

In pressurised aircraft the windows are very definitely part of the structure and have been developed into a standard form. In the great majority of designs there are three separate layers – a load-carrying layer, a back-up load carrier in case the first one cracks, and an inside plastic shield to prevent scratching by passengers. The load carriers will probably be glass, either heat-treated toughened glass like ovenware or laminated three-ply glass. (Again we have trade names for these products which have been adopted into the lanquage, in this case 'Pyrex' and 'Triplex'.)

The most usual transparent materials are acrylics (Perspex, Plexiglass, etc.) in sheet form. Although there is an inflammability problem with acrylic, it is not so serious as to exclude the use of the material. However there is now a newer range of transparent plastics called polycarbonates which are more fire retardant.

6.2.7

All the materials mentioned in the above paragraphs have their own virtues and values. It is sometimes thought necessary to make a comparison of materials by what is known as their strength/weight ratio. At first glance this appears simple and indeed for a straightforward tension member it probably is, but for any more complex piece of design there will be so many factors to consider that the material will almost certainly be selected on the basis of some other criterion. If we discuss the selection of material on a component to component basis then probably the only sure method is to make complete designs for each component in the alternatives and only then select the lightest design. As there is seldom time available for such an exercise then perhaps the strength to weight ratio may have some significance as a guide, but it can only be used with caution.

6.3 Cost as a Property of Material

Materials are selected to perform particular functions on the evidence of their properties and hence their suitability. When the function is structural then the major material property to be considered is strength. However, if strength considerations permit the possible use of several different materials, then other factors begin to affect the selection. Quite clearly one of these factors is cost. Unfortunately, while it is possible to list the cost of materials in terms of price per unit weight, this is virtually no help in making a decision.

Cost is more properly associated with the finished component, that is, bringing into account manufacturing costs which are fairly easy to assess, although even this assessment may be affected by quantity as well as by the difficulty of working the material. If a component is to be produced in very small quantities the tooling and design/drawing costs will create a different situation from that associated with large quantities. In these circumstances it may, for example, be cheaper to make a single shelf out of sandwich board with ready-made rigidity, rather than make a sheet metal shelf with stiffeners and formed edges. The sheet metal material would be cheaper than the sandwich board but the design/drawing time for the sheet metal construction would be much greater. Major companies now run special departments to assist the designer with decision making in these areas but in most design organisations, especially when dealing with the mass of small detail parts, very little conscious consideration is given to these factors. Unfortunately, two other large problems affecting the whole aircraft industry are becoming larger and are now clearly influencing material selection. One is replacement cost of life expired (worn out) components bearing in mind that this must include labour and inspection costs and the other is operating cost in terms of fuel expended in carrying weight during the life of the aircraft. At the time of writing:

the price of aluminium alloy is	= $ 6.00/kg
the price of titanium alloy is	= $44.00/kg
the price of turbine fuel is	= $ 0.20/kg
the average all up weight of a Boeing 747 half-way between London and New York	= 280 000 kg
the average cruising fuel consumption	= 11 000 kg/h
aircraft operating life assuming 20 years at 4000 hours per year	= 80 000 h
$\therefore$ total fuel consumed during the life of aircraft	= (80 000 $\times$ 11 000) kg = 880 000 000 kg
total cost of fuel during life of aircraft	= (880 $\times$ 10^6 $\times$ 0.20) = $176 000 000
$\therefore$ the fuel cost of carrying 1 kg during life of aircraft	= (176 $\times$ 10^6) $\div$ (280 $\times$ 10^3) = $628

Beside this large figure, the additional cost of specifying, for example, titanium instead of aluminium (where all other factors are equal) is insignificant.

(We should be aware that the calculation above is very crude and only intended to indicate the direction of an argument. One opposing argument is that with investment interest rate running at about 20%, a dollar spent today would be worth $15.40 in 15 years.)

In spite of this arithmetic, aircraft have to be sold and airline buyers still tend to look at initial price, so that although material cost is important, the relative costs of different materials may still prevent the lightest solution to a design problem being adopted.

6.4 Heat Treatment

6.4.1

The heat treatment of steel is not now normally carried out by the aircraft manufacturers because of the complex heat treatment requirements of these high strength steels. Nevertheless, it is useful to have some knowledge of the terms and expressions used.

Heat treatment changes the characteristics of a steel, changing one desirable feature for another. For instance, a tough springy alloy may be softened so that it can be bent and then rehardened to its former properties. In the carbon steels, the carbon is held in the alloy either as *pearlite* or *cementite* and it is the difference between these two states that determines whether the steel is soft or hard. When the material is red hot the carbon is in cementite state and if allowed to cool slowly the carbon changes to pearlite and the material is soft. (Used here, 'soft' is a comparative term because, even in this condition, the high-carbon and alloy steels are still difficult to work.) If the material is cooled rapidly from red hot by quenching in oil or in water, then the carbon is trapped in its cementite form and the steel is strong, hard and brittle. This brittleness is subsequently reduced in exchange for some reduction of strength and hardness by heating to a lower temperature and allowing some of the carbon to change its form. The three conditions described above are entitled:

annealed (or sometimes normalised)
hardened and
tempered

but for the structures engineer the names, with some idea of their meanings, are more important than the processes because specifications often use these words in their titles. For example, a steel might be described as 'annealed steel wire for oil hardened and tempered springs'. Also, the terms have been carried over into the description of light alloy heat treatments.

Maraging steels are heat treated in rather a different way in that some part of one of the alloying elements is persuaded to break its bond with the other constituents, leaving exactly the correct amount to give the material properties desired. When the element in question is fully dispersed through the material it is said to be in solution, and while it is changing it is said to be precipitating. The heat treatments which induce the change from one condition to another are thus called *solution treatment* and *precipitation treatment*.

6.4.2

The heat treatment of aluminium alloys follows roughly the same pattern as that for steels. That is, an alloy can be in the annealed or soft state and then be made progressively stronger by solution treatment and precipitation treatment which are roughly equivalent to the tempering process for steel. Most aluminium alloys work harden like copper, and this property is exploited in some workshop processes (see chapter 7). Because of work hardening, light alloy sheet comes out of the rolling mill in a condition which might be described as 'unnatural' and before being supplied to the aircraft manufacturer it is heated and cooled to change its characteristics by dispersing the crystals of the alloying elements (like copper) evenly through the material. This particular solution treatment or *normalising* process is carried out by soaking the material in a bath of melted salt, or by baking it in a hot air oven at approximately 450°C for about half an hour and then quenching it in water. (More exact figures are shown in table 6.4.) An interesting property of most light alloys is that immediately after normalising, and for about half an hour longer, the metal is soft. This property is useful for the manufacturing processes of bending and forming. After about an hour the solution-treated material starts to age harden. This is an important process because it continues to some extent throughout the life of the metal and components made from it. Precipitation treatment is artificial or accelerated *ageing*.

The various stages of heat treatment are called the 'conditions' of the alloy and are listed and coded by the Standards Bureaux of the major manufacturing countries.

British and American codes are shown in table 6.4 with space for noting other equivalents. In itself, this information is not very important to the engineer, but it is a useful reference when considering the specifications of common aircraft materials, because variations in treatment conditions may result in considerable changes of strength.

TABLE 6.4 Heat treatment designations for light alloy

Ref. no. (see also table 6.2)	Solution treatment temperature (°C)	Precipitation temperature and time	British designation	USA designation	Other designations
2014	502	None		−T4	
L.164 L.102	505±5	None	TB		
2014	502	160°C for 18 h		−T6	
L.165	505± 5	175°C for 10 h	TF		
2024	493			−T4	
L.109	495± 5	None	TB		
QQ–A–250/11 L.111 L.113 L.114	510–540	180°C for 8 h	TF		
7075	482	121°C for 24 h		−T6	
L.88	460± 10	135°C for 12 h	TF		

NOTES: This table illustrates some of the difficulty in being certain of exact equivalence in similar specifications. For example 2014–T4 and L.164 are similar materials (see table 6.2) but the difference in solution treatment temperature (502°C to a possible 510°C) might make a difference.

6.5 Reference Numbers for Materials

The quantity of different variants in the materials used in aircraft construction is enormous and totally bewildering. All of the major industrial countries have their own alloys, and to compound the confusion each has their own unique reference numbering system. Indeed some countries have more than one system. The United Kingdom has at least two systems for aerospace quality materials and the USA has at least three for aluminium alone. One of the American systems has, at least in theory, been adopted as an international standard and some aluminium suppliers refer to it in their brochures. Within this system which was designed by the Aluminum Association, alloys are allocated a four digit code number, followed by a code which identifies the heat treatment condition. The pattern of the whole number then looks like this:

2024–T3

The first digit indicates the principal alloying element and there are a total of seven groups of which only the 1000, 2000, 6000 and 7000 series are of much interest to the aircraft structures engineer. The code is as follows:

    1000 series   — pure aluminium (99% pure)
    2000 series   — copper alloys
    6000 series   — magnesium alloys
    7000 series   — zinc alloys

(In para. 6.2.3 we discussed three groups of alloys. The first group (Duralumin) is in the 2000 series and the third group is in the 7000 series. The middle group (known in the UK as Y-alloy derivatives) is mainly in the 2000 series but variants appear in the 6000 series.)

The second digit of the code denotes a revision to the original alloy (thus 2618 is the sixth modification of an alloy originally called 2018).

The third and fourth digits are index numbers.

Following the dash, the principal codes associated with aircraft materials are:

    –0   annealed
    –T3   solution treated and then cold worked (i.e. flattened or straightened) where
          this improves mechanical properties
    –T4   solution treated and naturally aged
    –T6   solution treated and artificially aged

There are a good many of these heat treatment codes to cover complex conditions. For instance, –T6511 is a subvariant of the –T6 state indicating that the product has been stress relieved by stretching and subsequently straightened. A full list of heat treatments, alloys, designations, etc., will be found in reference 1 (para. 6.6) but of the very many alloys and variants available, those shown in table 6.2 constitute the major interest of aircraft structures engineers.

A major snag with this otherwise worthy system is that the principal American reference numbering system for steels also uses a four digit code which is totally unrelated to the aluminium code. Fortunately the main aircraft steels fall into groups 3000 and 4000 which do not conflict with the aircraft aluminiums. Also the number is usually preceded by the letters SAE (Society of Automotive Engineers) but caution is necessary, for example, a screw made of SAE.2330 is steel not aluminium–copper alloy.

Other numbering systems used in the USA are Federal Specifications, Military Specifications and Aerospace Material Specifications and they look like this (in order):

$$QQ-A-367$$
$$MIL-A-22771$$
$$AMS.4127$$

These systems also cover materials other than metals. In fact there are MIL specifications covering every imaginable aircraft material including sewing thread, and for example an adhesive product may meet Federal Spec. MM-A-132, Type 1, Class 3, and MIL-A-5090-D, Type 1.

A further word of caution is necessary here, the Aluminum Association's numbers identify the material but the Federal and Military Specifications list the *requirements* for the material. This is really a quality control matter and the British system, which is discussed below, recognised this from the beginning (see also chapter 10). The official specifications, by listing minimum performance standards and by requiring that those standards should be demonstrated by test, provide a clear level of confidence for the designer and eventual user and an unequivocal basis against which the material manufacturer can quote prices.

The British industry uses two systems which do not overlap. A material which in the first place has been developed by private industry has its characteristics listed by a government department (Directorate of Technical Development) in the form of a specification. This specification carries a number which looks like this:

$$DTD.610$$

The material, which is now called by its specification number, is used and developed and if it becomes well established it is adopted into the British Standards Institution's list of aircraft materials and given a number which looks like this:

$$L.72$$

In fact the complete number would be BS.3 L.72 where the '3' signifies the third issue of the specification. (This incidentally brings up another minor difference between British and American practice. The British number the issues of a document and the Americans number the revisions. Thus when a document suffers its first change, if it is British it becomes issue B or issue 2 and if it is American it becomes revision A or revision 1.)

The strength and weakness of the British specification numbering system is that the number is just a number with no hint of indexing about it. For instance, L.102 is a bar material and its published specification is complete; from it we can find the heat treatment condition, the strength and almost anything else we could want to know about this particular bar stock. The same applies to a sheet material called L.164. What we cannot quickly recognise just from the numbers is that they are related and are both 2014 alloy.

(The random nature of these numbers is also illustrated by the history of the specifications for Dural bar. First issued in 1924 as DTD.18, the specification reached its third issue as DTD.18C and then became successively L.1, through to 6 L.1, L.39, L.64 and L.102.)

DTD and British Standard numbers also cover other materials. The DTD numbers do not change their pattern for other materials but in the British Standards steels (for instance) come into an 'S' group, e.g.

$$S.80$$

is a stainless steel.

Tubes can be confusing, L.114 is a light alloy but so was T.4.

As we said at the beginning of this paragraph, the numbering situation is very confusing but structures designers and stressmen quickly become familiar with the materials favoured by the company for whom they are working.

6.6   References

Aluminum Standards and Data
The Aluminum Association
750, Third Avenue
New York. NY 10017

The Properties of Aluminium and its Alloys
The Aluminium Federation
Broadway House
Calthorpe Road
Birmingham. B15 1TN

CHAPTER 7

Processes

7.1 Introduction

7.1.1

The range of manufacturing processes used by the aviation industry is so wide that to give only an indication of its scope in the same way that this book gives an introduction to structures would need a companion volume. We will therefore mention only a few of the basic processes, and show some photographs and diagrams of the machinery used.

7.1.2

The workshop processes of joining components by riveting, etc., is important in the construction of aircraft and we take this opportunity of showing how bolts and rivets are identified by stores reference numbers, as well as showing something of the range of fastenings available.

7.2 Manufacturing

7.2.1

When the aviation industry wants to specify a single piece of an engineering assembly it will refer to a piece part, a part, an item, a detail, a component or sometimes even to 'a dash number'. An assembly will have an associated parts list, bill of materials or schedule of parts (or some permutation of these words) which itemises the components of the assembly, notes the quantity of each that is required, names each item (or states its 'nomenclature'), and gives the part number or stores reference of each item. This number is also (usually) the reference number of the engineering drawing of the item. Often a drawing will give information on a range of similar parts, especially in the type of situation commonly found in the aircraft industry where parts are handed (that is parts installed on opposite sides of the aircraft are 'mirror images' of each other). In this case the identifying part number will usually be the drawing reference number plus a suffix or dash number. For example a port or left hand wing may be shown on drawing number 28503 and have a reference assembly number 28503–1 while the starboard or right hand wing will be 28503–2. The number of variants of this numbering system is the same as the number of companies in the aerospace industry, but the principles are fairly general.

7.2.2

The majority of aircraft components are either sheet metal parts (self-explanatory) or machined parts which may be made from bar stock, extrusion (see para. 6.1.2), forgings, castings or occasionally tube or plate. The exceptions are forgings, castings or lengths of extrusion which are used as supplied or which receive some minor work such as drilling (see fig. 7.1).

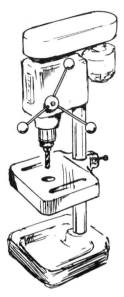

Fig. 7.1   Drilling machine

7.2.3

Machined parts are those which have been worked on a machine tool. The most common machine tool after the drilling machine is the lathe (see fig. 7.2). Lathes produce parts which have circular sections, such as bolts, wheels, bearing bushes, pistons, etc.

In aircraft production, however, the milling machine has probably overtaken the lathe as the most important machine tool (see fig. 7.3). The milling process makes smooth flat faces, grooves, shaped recesses or slots. It does this by cutting chips of metal from a surface by rotating a tool with one or more cutting teeth and driving it through the metal to be removed. By variations on this process some very complicated shapes can be produced (see fig. 9.12).

7.2.4

Although sheet metal parts are also worked on machine tools, these are in a separate range from those mentioned in para. 7.2.3. The tools for working sheet metal are enlarged and power-operated versions of the hand tools used by sheet metal workers for hundreds of years. With the exception of hole drilling, which is a very basic metal working operation, there are only two operations which can be performed on sheet metal – it can be cut and it can be formed. (Note here that the difference between 'sheet' and 'plate' is usually taken to be that material over 0.25 in or 6 mm is plate, but the 'sheet' referred to in this paragraph is less than 0.15 in or 3 mm.)

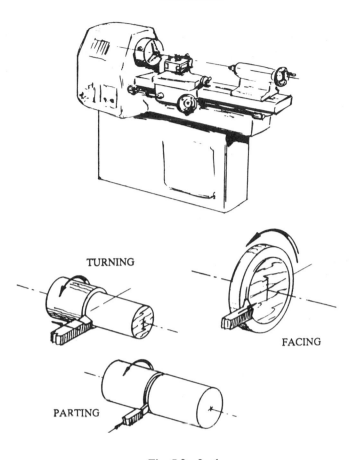

TURNING

FACING

PARTING

Fig. 7.2   Lathe

7.2.5

Cutting can be done in one of three ways, by guillotine, blanking or routing. The guillotine or shear (fig. 7.4) operated either by foot pedal or power, makes long straight cuts (4–10 ft dependent on machine size).

Blanking is a punch and die process. Any irregular shape can be cut out as a hole in a thick (comparatively) steel plate to make the die. A matching piece of similar material (the punch) made so that it will just pass through the hole, is mounted above the die in a press. When a piece of thin sheet is placed between punch and die the punch is pressed down and cuts one piece of matching shaped sheet. The process is very versatile and can reproduce accurate components, at a very high speed if necessary, from subminiature size to dimensions limited only by the power of the press. The tools (that is, the punches and dies) are expensive but the labour cost of the finished component is low, therefore this method is economical for long production runs with the runs needing to be longer as the component size (and therefore tool size) increases. Presses vary enormously in appearance and even with their exact function. A simple hand-operated machine, called a fly press, is shown in fig. 7.5, but hydraulic power presses of 500 tonne capacity are quite common equipment.

Routing also reproduces irregular sheet metal shapes. In this method a special milling cutter run at very high speed (18 000–24 000 rpm) cuts through a pack of five to

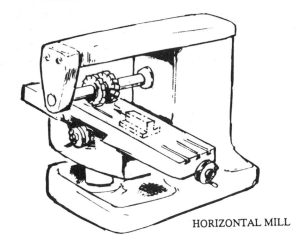

HORIZONTAL MILL

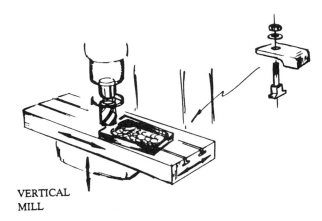

VERTICAL
MILL

Fig. 7.3   Milling machines

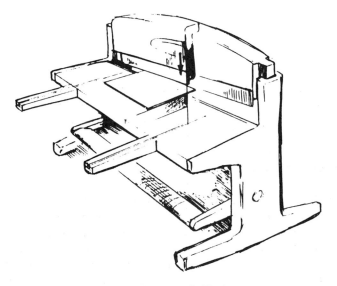

Fig. 7.4   Guillotine

ten sheets of material. The slot, which may be about $\frac{1}{4}$ in wide, can be of any length and follow any path guided by a template or model of the shape required. The tools are much cheaper than blanking tools and there is virtually no upper size limit, but the accuracy of profile is rather dependent on the skill of the operator and the labour cost per part is high. The name 'routing' is sometimes given to the milling process of making shaped recesses in plate material.

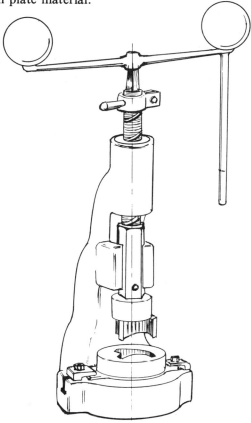

Fig. 7.5   Fly press

7.2.6

Forming sheet metal is probably the major activity of aircraft production. Straight bending of flanges or single curvature skin panels, such as those on the parallel central part of a fuselage or the surface of a conventional wing, is very easy. Forming parts with double curvature, such as the skin panels at the nose of an aircraft or the lips around the flanged lightening holes (see fig. 9.2), is considerably more difficult, but the processes are well understood in the industry and seldom cause major problems. There are some further comments on the design aspects of forming in chapter 9. Bending or folding of straight flanges is carried out in a folding machine (fig. 7.6) or a press brake (fig. 7.7). The latter machine can also be set up for blanking or piercing (i.e. blanking small holes). Double curvature forming may be carried out in a press with punch and die tools similar to those described in para. 7.2.5, but arranged with sufficient clearance so that the sheet is not cut between the two parts of the tool but is formed to the shape of the punch. Again such tools are very expensive and only justified by a long production run or absence of an alternative method.

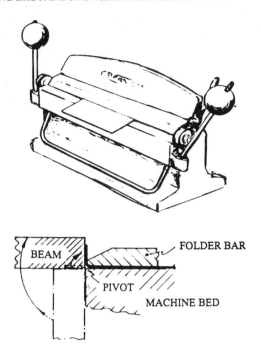

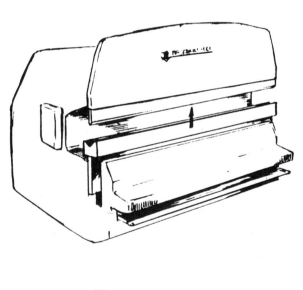

Fig. 7.6   Folding machine

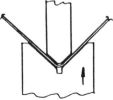

Fig. 7.7   Press brake

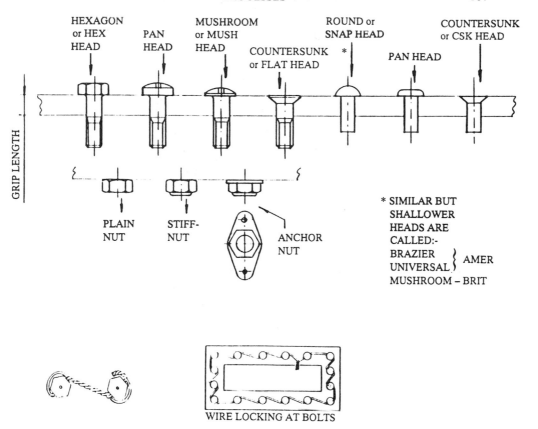

* SIMILAR BUT
SHALLOWER
HEADS ARE
CALLED:-
BRAZIER
UNIVERSAL } AMER
MUSHROOM – BRIT

WIRE LOCKING AT BOLTS

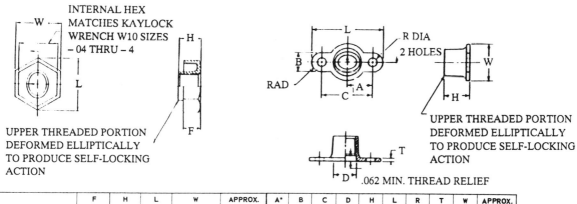

INTERNAL HEX
MATCHES KAYLOCK
WRENCH W10 SIZES
– 04 THRU – 4

UPPER THREADED PORTION
DEFORMED ELLIPTICALLY
TO PRODUCE SELF-LOCKING
ACTION

R DIA
2 HOLES

UPPER THREADED PORTION
DEFORMED ELLIPTICALLY
TO PRODUCE SELF-LOCKING
ACTION

.062 MIN. THREAD RELIEF

THREAD (MIL-S-8879)	F (REF)	H (MAX)	L (REF)	W	APPROX. WT. LBS./100	A* ± .005	B (REF)	C ± .002	D (MIN)	H (MAX)	L (MAX)	R + .005 − .000	T (REF)	W (MAX)	APPROX. WT. LBS./100
.1900−32 UNJF−3B	.143	.188	.433	.377−.365	.28	.250	.215	.500	.194	.220	.724	.098	.032	.328	.20
.2500−28 UNJF−3B	.164	.219	.505	.439−.430	.43	.281	.215	.562	.254	.281	.786	.098	.040	.414	.35

KAYLOCK SELF-LOCKING FASTENERS

Fig. 7.8   Some general information on fasteners (*Courtesy of Kaynar (UK) Ltd*)

A similar but cheaper, and in some ways more adaptable, version is the *rubber press* method. The punch is still required but is mounted on the bottom plate of the press and instead of a die the top plate carries a rectangular block of rubber contained inside a strong box or fence with the underside open. As the top and bottom press plates are squeezed together the rubber moulds itself and wraps the sheet metal around the punch. Punches (or form blocks) can be made of materials which are more easily worked than the steels used for matched punch and die sets, and the rubber (or sometimes now a plastic called polyurethane) is long lasting. Purpose built rubber presses are very large and may work at several thousands of tonnes load.

The process of forming sheet metal depends to some extent on the property of material called work hardening (see para. 6.2.2) and luckily aluminium, the major material of aircraft structures, has good work hardening characteristics. To illustrate this point imagine a component made of sheet aluminium and shaped like a deep dish or saucer. The manufacturing tools required will be a punch and die which respectively fit the inside and the outside of the finished component. Also required is a clamping ring or pressure plate to hold the unpressed metal blank in position over the die ready to receive the punch. As the punch is pressed down it starts to stretch the sheet into the hollow of the die, because the clamping ring (intentionally) prevents material being drawn in from the sides. Now as the material is being stretched it must get thinner and wherever it is thinnest we would expect it to be weakest and therefore to stretch more and so progressively become thinner in one small area. However, because of work hardening this is not the case. The initial small stretch 'works' the material improving the strength of the sheet locally, transferring the deformation to a new point and so by progressive action ensures that stretching takes place evenly over the whole area of the dish.

7.3   Jointing

7.3.1

Welding is the process whereby components made of metal or some plastics are joined by being made so hot in an area local to the joint, that the material of both components melts and runs together so that when the heat is removed the material is continuous across the joint. Usually, to compensate for gaps between the joint faces, a filler of similar material is melted into the weld. There are two other similar processes, *soldering and brazing*, in which the components being joined are not melted but are made hot enough to fuse a softer material such as tin or brass (note – brazing uses brass) which runs between them and acts like an adhesive.

Welding is a very useful solution to rather too many design problems in the sense that the designer may take the easy road out of his own difficulties and specify welding in place of some other process, thereby passing the problem to his colleagues in the production departments. Their difficulties lie in the fact that welding requires highly skilled labour and is extremely difficult to inspect for deficiencies.

More advanced welding techniques using lasers or electron beams are interesting for some specialised work, but from the structures viewpoint welding is not a process to be used too freely.

7.3.2

Adhesive bonding of metal structures is widely used by some aircraft manufacturers. All designers of airframe structures use adhesives for glass-reinforced plastic components and sandwich structures in various combinations of wood, paper, plastic and metal, but in the bonding of stringers to skins and other metal to metal joints in primary structure it is not so widely used. The De Havilland Comet and the Fokker F.27, products of the early 1950s, both used Redux bonding in their main structures. In both aircraft it was entirely successful and reliable and the two manufacturers have continued to use the techniques in their later designs.

One structural disadvantage of riveted construction is that every rivet hole is potentially the start of a tear or a crack, so the apparently simple drilled hole in fact requires care in its preparation, and an additional reserve of strength in its surrounding structure with the consequent weight disadvantage. (This would be referred to as a *weight loss*.) Bonded structures go a long way towards removing this particular problem and in many ways they are effectively similar to structures machined from solid stock material. Possibly even more effective is a mixture of machining and bonding to make the type of structure shown in figs. 9.10 and 9.11. A further major advantage with bonded structure is the excellent exterior finish produced.

There are some manufacturing problems. Cleanliness before bonding, application of pressure and heat during bonding and inspection of the joint after curing of the adhesive are all difficult. The provision of the necessary facilities should therefore be a matter for a company decision rather than a design office dictate.

7.3.3

Some of the design aspects of bolted and riveted joints are discussed in chapter 9. In this paragraph the patterns and types of rivets and bolts and some special fasteners are shown, together with notes on the methods of identifying various standard hardware items for design and drawing office purposes. Various shapes of rivet and bolt heads are shown in fig. 7.8 and the name description for a particular item would be in the following pattern:

Rivet, $5/32$ diam., csk head (read: rivet five thirty two diameter, countersunk)

Bolts are often named by their thread form:

Bolt, 10–32, hex. head (read: bolt ten thirty two hexagon head)

These names are only a general indication of the part, and the positive identification for specification or call out purposes is by a unique number. These numbers are in an even more bewildering profusion than the material specification numbers referred to in chapter 6.

Table 7.1 lists some of the major prefixes, and although we shall be referring to bolts and rivets the systems also embrace washers, hinges, clevis pins, etc., and many other items of hardware.

TABLE 7.1

Name of standards group or bureau	Item numbers prefixed by
American:	
Air Force–Navy Aeronautical Standards	AN
National Aircraft Standards	NAS
Military Aeronautical Standards	MS
British:	
British Standard Aircraft Part	$\left\{\begin{array}{l} SP \\ A \end{array}\right.$
Society of British Aerospace Companies Standard	AS
Aircraft General Standard	AGS

The prefix is followed by a number which identifies the style of bolt or rivet, that is its head shape or its particular function. In some cases the number also refers to size but this particular practice is being dropped in the more modern systems.

Examples:

(Amer.)	AN4	refers to a ¼ in diam. hex. head bolt
(Amer.)	AN365	refers to hex. nuts (but no particular size)
(Brit.)	A102	refers to hex. head bolts
(Brit.)	SP85	refers to mushroom head rivets
(Amer.)	MS27039	refers to pan head screws

The third part of the number completes the unique reference, so that dependent upon the amount of information given by the number so far, the third part must indicate at least:

diameter   (and/or thread form)
material
length

Examples:

(Amer.)	AN4CH10	is a ¼–28 UNF threaded hex. head bolt	(AN4)
		made of corrosion-resistant steel	(C)
		head and shank drilled for locking wire	(H)
		grip length ⁹⁄₁₆ in	(10)
(Brit.)	A102.5E	is a hex. head bolt made of high tensile steel	(A102)
		with a grip length of 0.5 in	(5)
		and ¼–28 UNF thread diam.	(E)
(Brit.)	SP85.410	is a mushroom head rivet made of aluminium alloy to specification BS.L86	(SP85)
		with a diameter of ⅛ in	(4)
		and length of ⅝ in	(10)

(Amer.)   MS20426DD3-12   is a csk. head rivet   (MS20426)
                          made of aluminium alloy 2024–T4 (DD)
                          $\frac{3}{32}$ in diameter   (3)
                          $\frac{3}{4}$ in long ($1\frac{2}{16}$ in)   (12)

As the reader will see, the standards organisations have devised some complex number codes, but with practice the systems become more readily understandable. For British readers, B.A.S. (aircraft components) Ltd publish a concise catalogue, and for American readers, information from Litton Fastening Systems Inc. is equally helpful. (The addresses for both companies are in para. 7.4.)

7.4   Reference Addresses

B.A.S. (aircraft components) Ltd
45 Vyner Street
London E2 9DQ

Litton Fastening Systems Inc.
3969, Paramount Blvd
Lakewood
California
90712

CHAPTER 8

Corrosion and Protective Treatments

8.1 Nature of Corrosion

8.1.1

Almost all construction materials seem to deteriorate. Buildings made by the Romans two thousand years ago have almost all disappeared, worn away by a natural deterioration which nobody troubled to prevent. Most materials try to go back to the form they were in before they became building material. There are exceptions like gold, glass and some plastics but aircraft materials in general either rot or corrode. The earliest aircraft only needed to have short lives; either they crashed or they became obsolete before they had time to wear out. The first deterioration troubles became evident in the 1920s in wooden aircraft when water found its way into glued joints and lay trapped in undrained corners. Indeed a major cause of the aircraft industry's swing to all-metal aircaft was an expectation of longer life. In general metal aircraft are very free from deterioration by corrosion but only because their designers and makers are aware of the problems involved and have taken a great deal of care to avoid them.

8.1.2

All metals can be chemically attacked and changed into salts of one type or another. The most familiar change of this type is the change of steel into rust which is iron oxide. Aluminium corrodes into a white powder. Some metals are recognised as being resistant to attack but by that we usually mean atmospheric attack and consequently some metals like stainless steel can catch us out and give unexpected problems.

8.1.3

Although there are many different ways in which metal can be attacked, for all practical purposes aircraft metals are damaged by an *electrochemical* process which takes place between the metal and another substance in the presence of an electrolyte. This happens so easily and the other substances are so common that really all that is needed to set up some corrosion is the electrolyte – water in small amounts is an excellent conductor. Having said that, we must make it clear that the problem varies in intensity. There is always trouble when two different metals are close to one another and we shall consider this situation in more detail below. Aircraft operating close to the sea have very severe problems. Naval aircraft and oil rig helicopters require special protection and the owners of the few remaining flying boats carry on a running battle against the effects of salt water which assist the metal to combine with atmospheric oxygen very quickly.

Electrochemical, *electrolytic* or *galvanic corrosion* is basically an action similar to that which takes place in an electric battery. Metal, electrolyte and another substance (which may be another metal) form an electric cell; current flows and chemical changes take place especially on the anode material which is the one with the higher electropotential. In an assembly, if dampness penetrated between an unprotected steel bolt and an aluminium alloy fitting, an electrochemical action would be set up and some of the aluminium would be changed to aluminium oxide. This 'cell' action can take place over very small areas of the same piece of metal as in the case of direct chemical attack (by acid or strong alkali), and it can even take place between the constituent metals of an alloy. Generally, corrosion needs a supply of oxygen and surface oxidisation is the common type of deterioration which we know well and know how to combat. Other forms of corrosion are fairly obvious like the deterioration of aluminium into aluminium sulphate under the action of battery acid. A form of corrosion which is extremely difficult to counter is that form called intergranular corrosion and its associated form, *stress corrosion*. As its name implies this attacks the interior of the metal, effectively treating each grain as though it were a separate component and spoiling its surface. This happens in spite of lack of oxygen, in an action which does not seem to be very well understood. It certainly is understood that the action is accelerated when components are under load, including the types of load induced by the manufacturing processes or heat treatment and known as residual stress. Unfortunately the corrosion has a knock-on type of effect in that the internal damage in the material is very similar to cracking, with the attendant sharp ends which are stress raisers, propagating in turn more cracks.

8.1.4

As with so many problems in the aircraft industry the causes are no worse than they are for other industries, but the effects are worse because we always have to push our materials, structures and indeed all our engineering, right to the limit of their capabilities. Consequently, when corrosion attacks a component it is not eating into fat, it is eating into the bone. Apart from intergranular corrosion which requires specialist equipment and expertise for its detection, most other corrosion is fairly easy to see (provided that it is not attacking a piece of structure which is hidden away from sight). Some corrosion like rust or copper oxide is coloured and makes telltale marks or stains. Surface corrosion has the advantage (if such a word can be used in connection with corrosion) that the oxide formed has a much greater volume than the metal from which it was formed. This phenomenon is well known to the owners of elderly motor cars when the product of corrosion under paintwork pushes up a paint blister out of all proportion to the size of the pitting in the metal. Similarly when aluminium corrodes, it produces a lot of powder which looks much worse than the damage justifies. None the less, the damage is there and advice must be sought from the aircraft manufacturer (via the maintenance manual) about how much damage is acceptable in a particular area before a component is scrapped. Strangely, although aluminium oxide is a pale grey to white powder quite unlike new aluminium, as a structure gets older and any internal paint starts to flake a little, the oxide is harder to see, or perhaps easier to overlook, than one would imagine.

8.2 Causes of Corrosion

8.2.1

In a sense the primary cause of corrosion is failure to provide a protection from atmospheric attack. However, this is a fairly obvious source of trouble and can be avoided most of the time, but there can be catches. For instance in a company where the usual practice is to use Alclad material (see para. 8.3.2), a sudden requirement which calls for the use of unclad sheet can lead people to leave the interiors of closed structures without protection. Some years ago manufacturers of crew and passenger seats were very fond of making their structures of welded tube, which is virtually impossible to protect on the inside (see para. 8.3.7). Unfortunately, where a structure which needs paint is left bare it is usually because the painter cannot reach it and this invariably also means that it cannot be seen so that any eventual corrosion is likely to remain undetected.

8.2.2

Another prime cause of corrosion must be failure, during routine maintenance, to carry out the manufacturer's instructions on protection; but this is a quality control matter (see chapter 10) and just outside the scope of this book.

8.2.3

The corrosion giving the most trouble to structures designers arises from contact between dissimilar metals. The situation is that one metal corrodes the other, but within certain very tight limits this can be acceptable, or at any rate tolerated. The existence of a degree of tolerance does sometimes allow a situation in which it is possible to interpose a metal separator between two non-compatible materials such that the middle material avoids seriously damaging either of the others. It is by this method that we can accept cadmium-plated steel fasteners in light alloy structures, but in most cases it is better to separate dissimilar metals with an insulator (plastic sheet, Tufnol or Paxalin preferably, or even jointing compound if the problem is only marginal). One of the traps for the unwary is the stainless steel (CRES) fastener which may need plating to avoid an unacceptable level of potential corrosion against some light alloys. Even different aluminium alloys or the same alloy at different tempers can cause trouble, although seldom so much that it cannot be avoided by the use of jointing compound. On the other hand, flying boats and oil rig helicopters sometimes suffer corrosion between light alloy skins and stringers, to the point where the growth of the aluminium oxide is sufficient to pull rivet heads through countersinks in the skin.

8.2.4

Stress corrosion was mentioned in para. 8.1.3 as a form of intergranular corrosion and this seems to affect the high strength zinc bearing light alloys more than some of the others. Following a remark earlier in this book (see para. 6.2.3) the reader will sense little enthusiasm for high zinc alloys, however, it is necessary to use them at times to save weight. In flat sheets or gently curved panels they are satisfactory, but in any situation where bending or heat treatment could possibly produce built-in unrelieved stresses they are prime candidates for stress corrosion. Another situation in which stressed metal can be attacked by intergranular corrosion is created under bolt heads which have been over-tightened. Designers should make sure that the workshops know the permissible torque loadings for fasteners in sensitive areas.

8.2.5

All riveted and bolted joints are liable to *fretting corrosion* (faying) which results from constant small relative movements between clamped faces. If the joint is in a structure there is almost certainly a load across the joint. If there is a load there is a movement (no stress without strain); so solutions to the fretting problem which involve clamping bolts up tighter are only wishful thinking. The solution is to separate the metal faces with jointing compound on permanent joints or with anti-faying lubricant on other joints.

8.2.6

The last type of corrosion which is particularly relevant to conventional aircraft structure is *shielding corrosion*. It appears where wood, furnishing materials or sound proofing hold a wet patch in contact with metal, and may at first glance seem similar to the type of corrosive situation in which areas of structure are compelled to lie in accumulated water. (It is not possible for instance to put drainage holes in the bottom of a pressurised fuselage, and toilet and galley spillage, plus condensation, can make a very corrosive combination.) This latter situation can easily be recognised and guarded against, unlike shielding corrosion in which the rate of corrosive attack is in fact greater than would be expected if the metal was totally immersed in water. Again there is a possible trap when using stainless steel in this situation. The particular property which makes stainless steel resistant to corrosion is an ability to instantly form for itself a thin protective skin. Unfortunately, this skin only forms in a well-ventilated situation. So in a situation of the type described above, the wet material breaks the surface skin of the stainless steel which then finds itself unable to immediately repair itself, therefore the steel is no longer stainless. Some CRES alloys in this situation can completely corrode away faster than unprotected mild steel. The only satisfactory solution is to insulate the metal from the water carrier in just the same way as one would insulate it from a dissimilar metal.

Even then there is still a snare which has deluded some eminent designers (especially in the automotive industry). In an effort to reduce maintenance down time, the dry lubricated bearing is a tempting proposition, but there are problems. Stainless shafts rotating in turned nylon bushes (moulded bushes with a very smooth bore are better) can suffer from shielding corrosion with the result that the shaft swells by corrosion and starts to squeak and bind.

8.3 Protection Against Corrosion

8.3.1

The classic and only satisfactory way of stopping a metal from returning to its earth is to cover it. Either paint it, plate it or allow it to produce its own oxide covering without disturbance. The last of these methods is the most debatable so we will discuss it first.

8.3.2

Most constructional metals and alloys start to corrode on their surface, and the initial oxide films so formed cover the parent metal and tend to inhibit it from further attack.

With some alloys this action can provide sufficient protection on its own as in the case of stainless steel (CRES) and titanium. Most oxide films are hard and tough but some are very thin and can be broken. Usually this is no cause for worry because a new protective skin immediately forms. If however a condition exists which prevents this, more serious corrosion may result. Designers who are relying on this form of protection should also ensure that components are well finished (polished), because small surface imperfections can enlarge into pits which in turn become stress raisers.

The major material of aircraft construction, aluminium. has, at least in its unalloyed state, a (comparatively) thick and stable or passive natural skin. It is so patient and gives so little trouble that we have to be careful not to take it too much for granted. In general the higher the strength of the alloy the more it is susceptible to corrosion, but even this problem is effectively defeated by the *Alclad* process. This elegantly simple but highly inventive idea is to coat the vulnerable high strength sheet with a very thin surface of non-vulnerable low strength pure aluminium. The coating is hot rolled on in the manufacturing process and forms an absolutely unpeelable bond. Some major aircraft constructors are satisfied that structures made from Alclad sheet are sufficiently protected, at least internally. Since other manufacturers paint everything (see below), there is obviously room for sensible debate. Unexpectedly, Alclad sheet is less vulnerable at cut edges and drilled holes than unclad sheet. This bonus is due to sacrificial protection by the unstressed aluminium in favour of the stressed aluminium alloy. This takes place as galvanic corrosion, acceptable because the loss of the soft pure aluminium takes place slowly and with no loss of strength to the structure. So although the Alclad process is in a very real sense a covering or coating protection, most of its efficiency derives from the self-healing passive oxide film on the aluminium.

8.3.3

One primary protection for steel parts is cadmium plating. This is a metallic electrodeposition process like decorative chrome plating or domestic silver plating on cutlery. As protection, cadmium plating works in at least three ways. Firstly, cadmium is a passive metal and offers good resistance to atmospheric attack. For this property alone it was considered a suitable protection when some vital steel structure components were exposed to the elements. Biplane tie wires were often left cadmium plated but unpainted. Secondly, cadmium plating acts as a sacrificial anode in a similar way to that described for aluminium cladding (para. 8.3.2). This means that cadmium-plated components are not sensitive to minor surface damage even when the plating is scratched right through. Thirdly, the cadmium acts as an intermediary between steel and aluminium as it has a level of electropotential acceptable to both.

Cadmium plating is a very thin film. Depending upon the exact application the film thickness usually varies between 0.0002 and 0.0006 inches (i.e. 0.005 to 0.015 mm) but as this is of the order of 0.001 inches on the diameter of bolts, etc., draughtsmen need to be careful when specifying manufacturing tolerances to state whether their dimensions are before or after plating.

The British specification for the whole process is DEF STAN 03-19 and the American specification is QQ–P–416; it is very much the responsibility of the designer to know what these specifications are giving him. In terms of the British specification it is, strictly speaking, necessary to stipulate a pre-treatment DTD 901 (degreasing), plus any subsequent treatments such as DTD 934 (which applies to high tensile and stainless steels); de-embrittlement which applies to springs; and passivation which applies fairly generally, and certainly when there may be a painting operation.

To assist draughtsmen, most design organisations provide a shorthand note, or some term which avoids repetitive delineation of exact processes. This does not, however, relieve the draughtsman of the responsibility of ensuring that he is asking for the correct processes. There is a principle here which applies over the whole field of aircraft and aircraft equipment design, and we can use the shorthand note 'Cad and Pass' to illustrate its importance. By putting such a note on his drawing, the draughtsman is stipulating a process – 'Cadmium plate to DEF STAN 03-19 preceded by . . .' etc., through to '. . . and passivate', and provided that he is prepared to accept the decision of the process department on the standard required, then his shorthand note is sufficient. If however, he is specifying a Class 1 component (see para. 5.8.2) then he must ensure that the process department understands that his note means nothing less than the full treatment. On the other hand, if the component involved is relatively insignificant, it would be wrong to waste office time, workshop cost and inspection effort on providing more than the minimum standard of finish. The information that the draughtsman is communicating should make this quite clear.

This general philosophy should apply to all the engineering department's work. Where quality requirements demand the best, the designer must be assured that he is going to get it. However, if there is no justification for expensive materials or processes or labour, then he should be equally certain that the draughtsman is not guilty of 'over-engineering' the design.

8.3.4

The standard finish for aluminium alloy components made of anything other than Alclad sheet is *anodising*, or more exactly, anodic oxidation. The current British specification reference for anodising is DEF STAN 03-24 and the American specification is MIL-A-8625. Anodising produces artificially, the tough oxide film on aluminium alloy which, as we already know from para. 8.3.2, exists naturally on pure aluminium. As far as Alclad is concerned, there is no reason why it should not be anodised, but the process is pointless unless the material is to be subsequently bonded by Redux (a trade name for an adhesive produced by Ciba–Geigy) or subjected to a similar operation, for which anodising provides an excellent preparation.

There are various processes under the general heading of anodising, including decorative finishes, and an abrasion-resistant hard anodising process, but for structural parts the chromic acid method is most widely used. This process, however, produces only a thin oxide film which is porous and easily damaged, and must therefore be sealed either by rinsing in boiling water or by being immediately painted (or at least primed, see below). Neither action is necessary if the part is to be bonded.

The remarks made in the previous paragraph, concerning the need to specify exactly what is required by the design, apply equally here. The draughtsman is in the business of communicating the designer's wishes to the workshop and he must make sure here, as much as in the more glamorous areas of aircraft design, that what he puts on paper carries the right message.

Some authorities suggest that it is possible to anodise welded or even riveted assemblies. The chromic acid process has the advantage that it shows up cracks in welds remarkably well, but even so the danger of trapping the fluids used in the process is too great a risk. The answer with riveted asemblies is to anodise or paint the components before riveting. The answer with welded parts is simply to avoid welding altogether; there is usually an alternative method that will give less trouble in manufacture, quality control and service.

(There are exceptions to this heavy dogma. Some automatic or machine-controlled welding and some electron beam or laser beam welding may justify its cost for some jobs, otherwise welding belongs with steel tube structures – back in history.)

8.3.5

An alternative to the full electrochemical immersion anodising process is the application of one of a number of proprietary pastes, such as Alocrom, which produce a protective film. These are especially useful on assemblies (or repairs) when the original anodising has been damaged perhaps by some machining process which could not be completed before assembly. However, these treatments must be used with caution as some, which at first sight appear entirely satisfactory, have to be painted over immediately.

8.3.6

There are a multitude of protective treatments which may apply to aircraft in general but the only remaining treatment which must be mentioned in relation to aircraft structure is paint. Even here there is a great variety of alternative finishes: cellulose, stove enamel, expoxy, acrylic, polyurethane with, no doubt, new ones appearing regularly. Paints are the universal protectors of structures and with the possible exception of parts made of Alclad sheet, structures need it. Paints usually contain inhibitors intended to ward off atmospheric corrosion chemically but essentially they provide a decorative covering less susceptible to deterioration than the structure, easier to replace when damaged and not contributing to strength.

When thinking of paint it is necessary to think of a whole paint scheme with three or two compatible elements:

(a) the primer, which is the protector for the material and is made to adhere to the material surface
(b) the undercoat, which is the opaque colour
(c) the top coat, a transparent decorative finish which also is hard or tough enough to protect the primer

Some aircraft schemes have been designed to save weight by dispensing with element (b). When selecting a scheme, check that if the paint is heat cured the heat can be applied safely and that the temperature does not approach the heat treatment temperature of the material being painted. Primers for metal surfaces are usually etch primers. The most universal contain zinc chromate which gives treated parts a distinctive yellow, or green–yellow finish which for some reason always looks stable and reliable. For interior parts which do not require a decorative finish, a good primer may be sufficient in itself.

Although some etch primers (especially the two-part synthetic resin types) are very tolerant of poor pre-priming surface preparation, degreasing by vapour bath for light alloy or mechanical abrading (sand blasting) for steel is almost essential. Decorative exterior paint schemes usually specify their own particular pre-treatment.

Some exterior paints are specifically applied to protect the strength of the structure from erosion by stone chippings thrown up from the runway or ice particles in the air. As with all other protective treatments, specification is the responsibility of the design office and must be complete.

8.3.7

There remain the components or parts of components which cannot be protected. The inside surfaces of tubular structures and the inside of holes reamed on assembly are typical examples. Unfortunately most manuals and handbooks are remarkably unhelpful on this subject and regrettably this book is no exception. Perhaps the most practical advice is to check that the part really cannot be reached. Maybe it can be reached by being dismantled after a trial fitting. In another case a coating of lanolin applied by any available method may be better than nothing, but the real answer is that if a part cannot be protected then the design is wrong.

8.4 Reference

Civil Aircraft Inspection Procedures, Part 1.
Civil Aviation Authority Printing and Publication Services
Greville House
37 Gratton Road
Cheltenham
Glouc. GL50 2BN

CHAPTER 9

Detail Design

9.1   Sheet Metal Components

9.1.1

The essence of detail design is understanding the limitations of the materials being used.

Although an increasingly large number of aircraft structural components are now machined from solid billets or thick plate, sheet aluminium alloy is still the major material of aircraft construction. Figures 9.1 to 9.5 show some typical components and it will be seen that apart from the skin panels, almost all the items are bent or formed in some way. When bending or 'folding' the structural alloys it is necessary to make a radius in the corner as shown in fig. 9.6. The actual radius required depends on the material and its heat treatment condition (see chapter 6). Within half an hour of normalising, most alloys can be bent to an inside bend radius equal to $2t$ (that is, twice the material thickness) but higher strength alloys may require a $5t$ internal radius or even more if the bend angle exceeds 90°. The penalty for trying to reduce the bend radius recommended by the material supplier is cracking or splitting of the metal on the outside of the bend. Unfortunately the requirement for a bend radius brings problems in its train which we shall discuss below and illustrate in fig. 9.15.

9.1.2

Bending sheet metal along a straight edge (also called folding) is a comparatively straightforward process but any variation from this simple bending causes manufacturing problems which are disproportionately severe. Most of the diagrams 9.1 to 9.5 show bends which have *joggles* (see fig. 9.4) and it is almost impossible to avoid their use in sheet metal structures but in addition to some reservations about their strength they are very difficult to make. The two small brackets in fig. 9.2 are a case in point; the smaller of the two is a simple job for the folding machine but the larger, being joggled, would require a special press tool and longer production labour time. These differences tend to disappear with larger quantities, but for small production runs joggling should be avoided where at all possible.

Bending flanges on curved members is also difficult. Fig. 9.7 shows a curved channel member. A component of this type would be made by pressing or hand beating a flat development over a block or tool specially made to the correct shape. Edge AA on the diagram is stretched during the forming process and can be made smooth, but the difficulty is that edge BB has to shrink and tends to fall into waves which are difficult to eliminate. Again bends of this type cannot be completely avoided but designers try to find alternatives where they can.

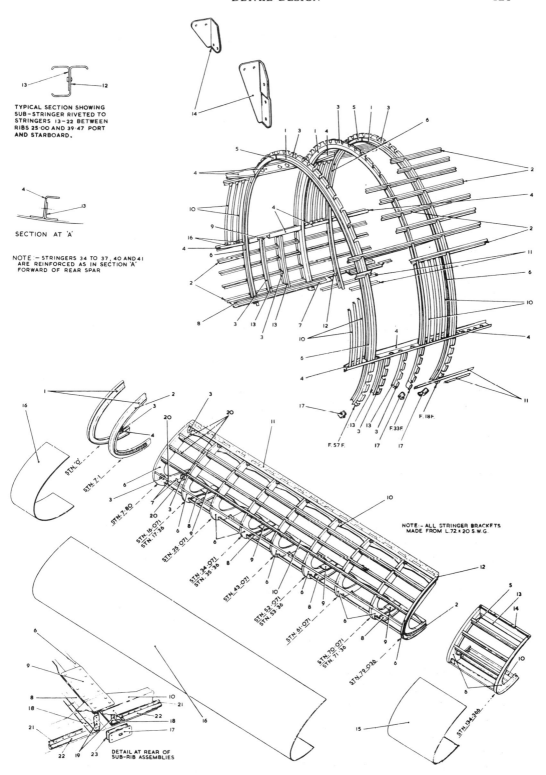

Fig. 9.1 Examples of detail design, B.Ae. 748 (*Courtesy of British Aerospace*)

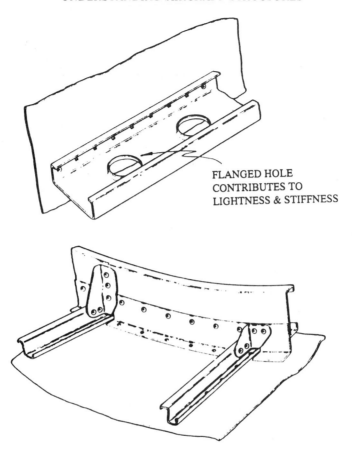

FLANGED HOLE
CONTRIBUTES TO
LIGHTNESS & STIFFNESS

Fig. 9.2   Typical sheet metal construction

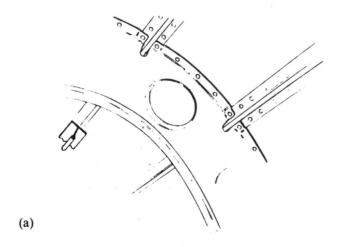

(a)

Fig. 9.3   ((a) *Courtesy of The De Havilland Aircraft of Canada Ltd*); (b) Boeing 747 (*Courtesy of The Boeing Commercial Airplane Company*); (c) B.Ae. 146 (*Courtesy of British Aerospace*)

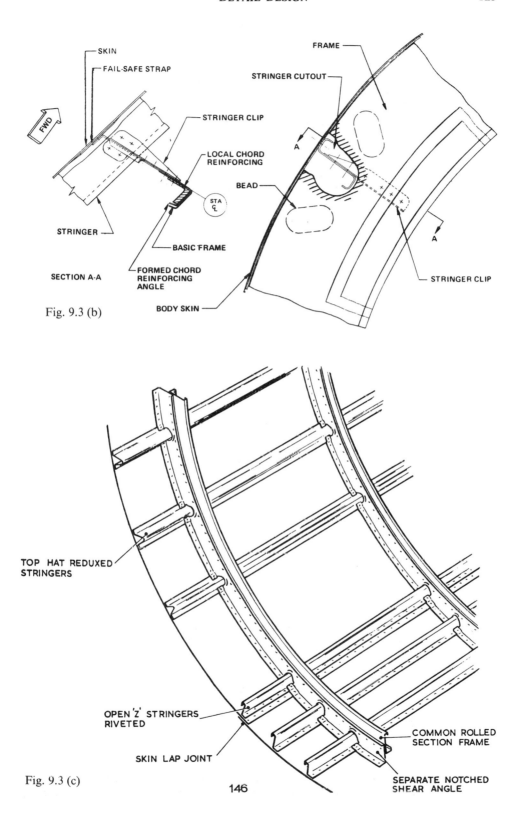

SKIN

FAIL-SAFE STRAP

FWD

STRINGER CLIP

LOCAL CHORD
REINFORCING

STRINGER

STA
₵

BASIC FRAME

SECTION A-A

FORMED CHORD
REINFORCING
ANGLE

Fig. 9.3 (b)

FRAME

STRINGER CUTOUT

A

BEAD

A

STRINGER CLIP

BODY SKIN

TOP HAT REDUXED
STRINGERS

OPEN 'Z' STRINGERS
RIVETED

SKIN LAP JOINT

COMMON ROLLED
SECTION FRAME

SEPARATE NOTCHED
SHEAR ANGLE

Fig. 9.3 (c)

146

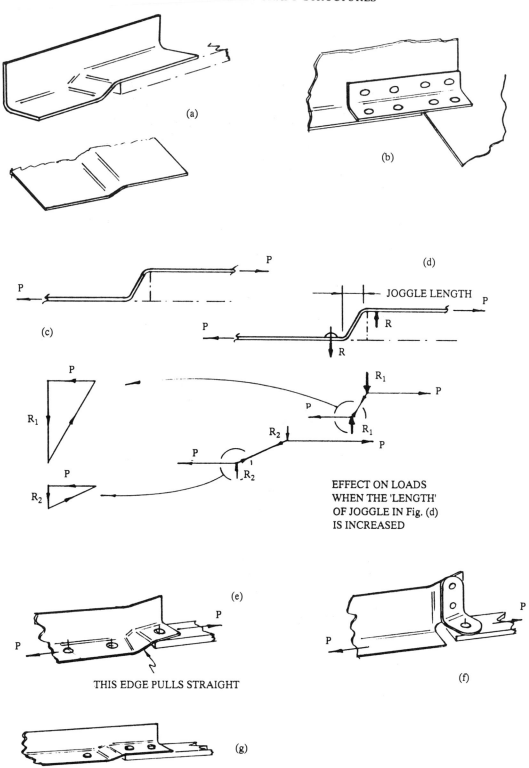

Fig. 9.4   Joggles

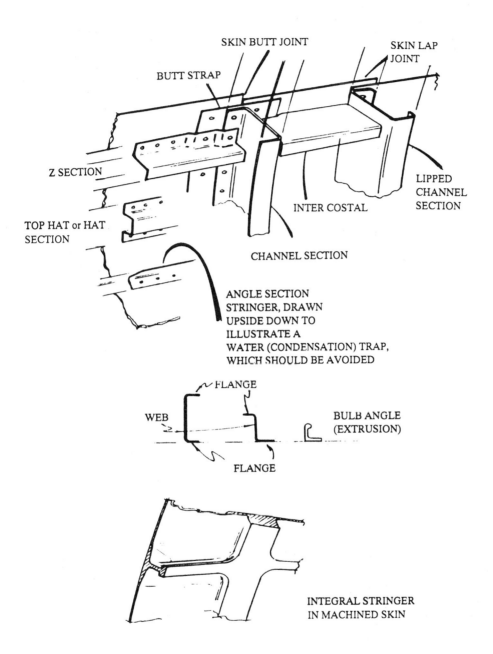

Fig. 9.5 Structural parts

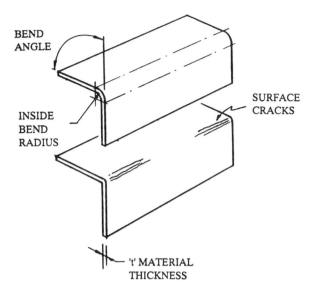

Fig. 9.6   Sheet metal bending

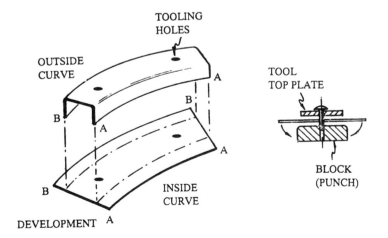

Fig. 9.7   Sheet metal forming

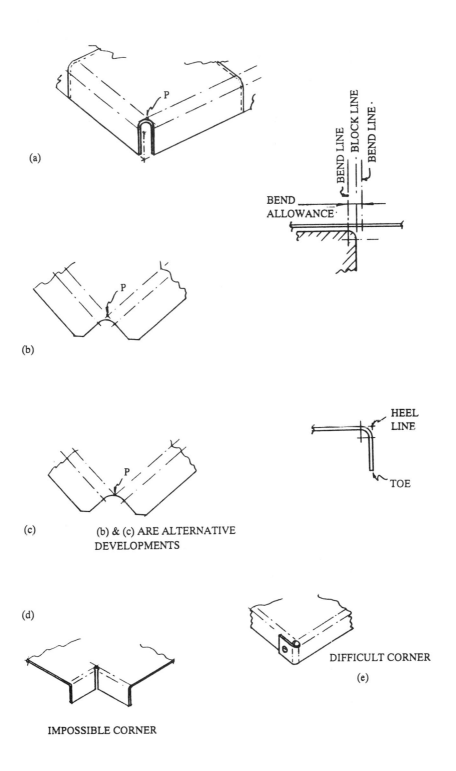

(a)

BEND LINE

BLOCK LINE

BEND LINE

BEND
ALLOWANCE

(b)

(c)   (b) & (c) ARE ALTERNATIVE
DEVELOPMENTS

HEEL
LINE

TOE

(d)

IMPOSSIBLE CORNER

DIFFICULT CORNER
(e)

Fig. 9.8 Sheet metal corners

9.1.3

The need for a substantial bend radius makes for difficulty in the corners of a certain type of sheet metal component or fitting. Fig. 9.8(a) illustrates the point and (b) and (c) show the effect on the development of the corner. (Developing a sheet metal part simply means laying it out flat.) For reasons which are discussed in para. 9.3 below, the shaded area in (b) cannot be cut out to leave a sharp corner at P; nevertheless some companies accept that development (b) is satisfactory while other companies require solution (c).

Fig. 9.8 also shows at (d) and (e) two design situations which crop up from time to time. Consideration of the developed shape will show that (d) is an impossible problem. Shape (e) is not impossible, but it is very difficult to make and should be avoided.

9.1.4

The more complex shapes in sheet metal are produced by pressing, which was mentioned in chapter 7 – workshop processes. The flanged holes in fig. 9.2 and the components with curved edges in fig. 9.1 would all be press formed. The methods of production are very versatile but designers have to consider how their components will be made and, as with all components, simplicity in the design of pressings is a virtue.

9.2   Machined Components and Large Forgings

9.2.1

Some aircraft structural components, such as undercarriages, are made almost exclusively of components machined out of solid bar material or forged blocks. This book is more concerned with the larger components such as fuselages and wings, and in recent years the percentage of such structures machined from solid stock has increased substantially. One of the major advantages of machined structure is the reduction of notching effects, which will be discussed in their relationship to riveted structure in para. 9.3. (Notching is the introduction of an abrupt change of section due to drilled holes, grooves, sharp-edged undercuts, etc.)

Fig. 9.5 and figs. 9.9–9.12 illustrate some typical machined structural parts. It can be seen in fig. 9.5 that there can be an immediate weight saving compared with riveted structures because of the elimination of overlaps at joints and stringer attachments. More careful and thoughtful design using adhesive bonding (as shown in figs.9.9 and 9.11) produces even more advantages. Savings on cost of materials, reduced machining costs by the use of extrusions, and the option of using very high strength alloys in selected locations are added to the more obvious advantages mentioned above.

As with almost all design options in aircraft engineering there are advantages and disadvantages. Repairs to machined structural components are much more difficult than repairs to sheet metal structures; initial tooling costs are very high; capital investment in machinery is very high; the cost of representative prototype test pieces is high and mitigates against risking early test failures which in turn suggests heavier than necessary structure. It is probable that for the next few years the larger companies dealing with advanced designs will increase their use of machined structures, while smaller aircraft companies facing commercial pressure for a lower initial selling price will continue to use the more traditional methods of construction.

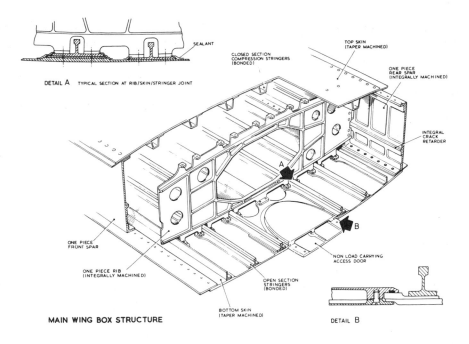

Fig. 9.9   Machined structural details (*Courtesy of British Aerospace*)

Fig. 9.10   Constructional details with machined components (*Courtesy of British Aerospace*)

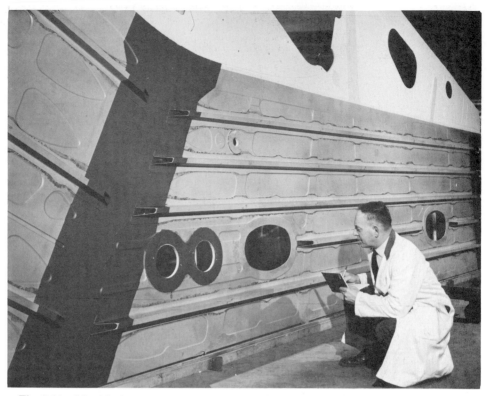

Fig. 9.11   Machined and bonded structure (*Courtesy of Ciba-Geigy, Bonded Structures Division, Cambridge*)

(a)

Fig. 9.12(a–c)   Structural components machined from forgings or heavy plate (*Courtesy of British Aerospace*)

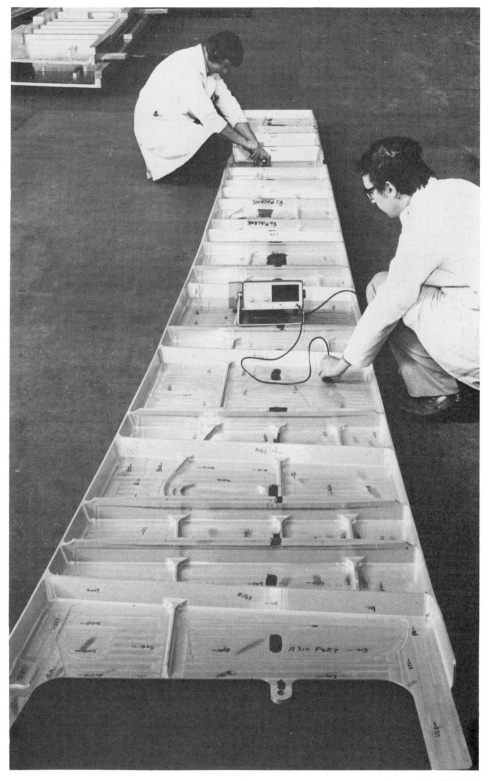

Fig. 9.12 (b)

Fig. 9.12(c)

9.3   Notching and Stress Raisers

9.3.1

Fig. 9.13 shows some examples of what is meant by *notching* which is perhaps the most dangerous trap of all for the aircraft structural designer. Apart from situations where ground staff have not fastened doors correctly, etc., and at some risk of sounding melodramatic, it is difficult to think of a structural failure leading to an aircraft accident in recent years that was not initiated by a *notch effect* which might have been avoided by better design. That is not to suggest undue criticism of the particular designers; all designers get caught by the ease with which unintentional *stress raisers* can be built in, and those who recognise and retrieve the situation before an expensive failure occurs know that they have been very lucky indeed.

9.3.2

The essential feature of notching is that because it is notched, a component is disproportionately more liable to breakage. The most common example of the phenomenon is its exploitation for glass cutting. Glass sheet which has been scored with a very light groove will break at a small fraction of the load which would be required to break unmarked sheet.

9.3.3

In fig. 9.13(b) the maximum stress in the material near the hole is much higher than the average stress which would be calculated by dividing the load P by the area shown shaded. Fig. 9.13(d) indicates the situation by showing the lines which represent an even stress distribution suddenly crowded together at the groove.

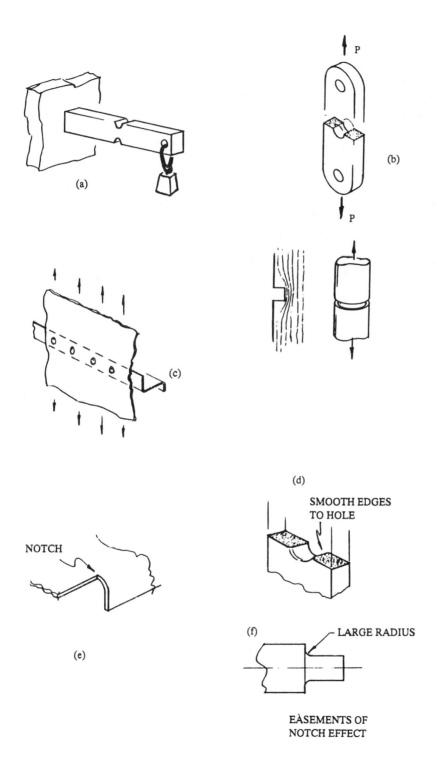

Fig. 9.13   Notches

Chobert

(a) A steel mandrel which has an opposite taper on the head, is drawn through from the tail end of the rivet, expanding the rivet tail around the rear side of the hole forming a shoulder.

(b & c) The mandrel continues to be pulled through the rivet symetrically expanding the rivet shank to fill the hole.

(d) This ensures the rivet has good bearing in the hole and a parallel bore is left in the rivet. The cycle is completed and the next rivet is ready to be placed into the prepared hole.

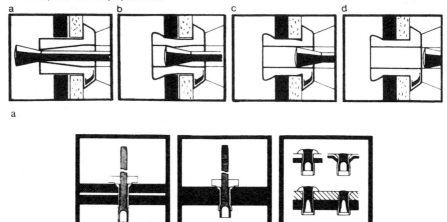

a

How the Avdel MBC Rivet system works

1. The MBC Rivet is loaded into the nose of the placing tool, and the tool applied to a prepared hole in the workpiece.

2. When the tool is actuated, jaws in the nose of the tool grip the rivet stem and exert an axial pull, drawing the stem through the rivet shell to give a high-clench joint and complete hole-fill.

3. The placing tool automatically shears the stem flush with the rivet head, and is mechanically locked into the rivet shell.

b

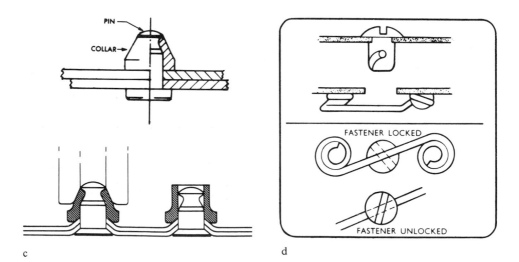

c   d

Fig. 9.14   (*Courtesy of Avdel Ltd* (a and b), *hi-Shear Corporation* (c), *Dzus Fastener, Europe Ltd* (d), *Cherry Rivet* (e), *Huck Fasteners* (f))

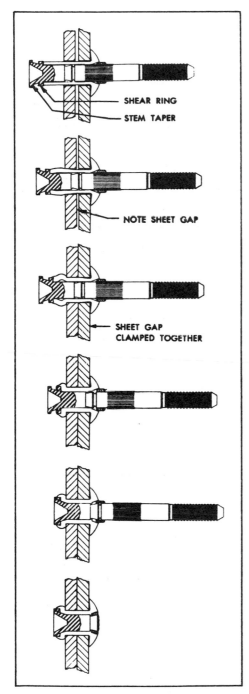

SHEAR RING
STEM TAPER

NOTE SHEET GAP

SHEET GAP
CLAMPED TOGETHER

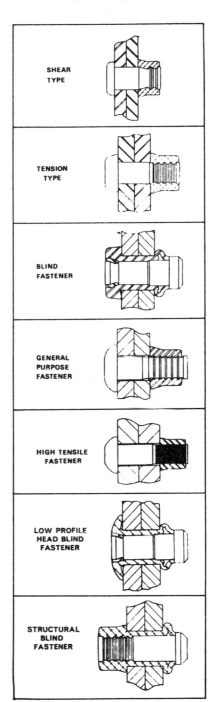

SHEAR
TYPE

TENSION
TYPE

BLIND
FASTENER

GENERAL
PURPOSE
FASTENER

HIGH TENSILE
FASTENER

LOW PROFILE
HEAD BLIND
FASTENER

STRUCTURAL
BLIND
FASTENER

e

f

A very common aircraft structure situation is shown in fig. 9.13(c). In this case if the rivet holes are $\frac{1}{8}$ in diameter and the rivets are spaced 1.0 in apart (3 mm diam.; 24 mm spacing) the actual maximum stress in the sheet will be approximately 2.6 times the nominal stress which would be expected from dividing the load by the sheet thickness and length.

9.3.4

Unfortunately neither structures nor machines can be designed entirely without notches. The best we can do is to recognise them, appreciate the type of effect they have and make allowances for them by lowering the nominal stress level. In the two examples mentioned in para. 9.3.3 we would probably have to make the bar and the skin thicker. (It should be noted here that using stronger light alloy material may not provide an answer because strong materials are often more susceptible to notching effects than softer, more ductile materials.)

The examples shown in fig. 9.13(f) illustrate simple precautions which must always be taken. Example (e) is also easily avoided (see fig. 9.8). Other problems which may crop up in a design will need their own solutions.

9.4 Rivets and Bolts

9.4.1

Rivets and nuts and bolts are part of a general group called *fasteners*. Under the same heading there is a large range of proprietary fasteners which offer particular advantages for special purposes, such as shear strength above the normal to be expected from a rivet, but with a lower installed weight than a bolt. Another special fastener acts in a similar way to a lightly loaded bolt, but is designed to be released by only a 90° twist with a screwdriver. This last example belongs to a subgroup called quarter turn fasteners and, although they do not have any relevance to major structure, they are used to retain light fairings, engine cowlings and access panels.

Some indication of the range of fasteners is shown in fig. 9.14 and notes on the methods of reference by number are given in chapter 7.

9.4.2

In chapter 7 reference was made to the *grip length* of bolts. When specifying bolt length it is important that the smooth shank of the bolt extends right through the hole so that no thread is against a bearing surface, so the grip length is the plain length and is approximately equal to the thickness of the materials being fastened together plus the thickness of any washer under the bolt head. For countersunk or flat head bolts the length is measured from the top of the head, and for other bolts from under the head (see fig. 7.8). In almost all cases, specifying the correct length of bolt shank will result in having a small amount of plain shank exposed past the hole so that a washer or washers will be needed under the nut to allow tightening.

9.4.3

When specifying the length of a rivet, allowance must be made for forming a head. This allowance is about $1\frac{1}{2}$ times the nominal rivet diameter but the design offices of the aircraft manufacturers have their own standards for this important dimension.

9.4.4

In a normal structural use of bolts the fastening is completed with a nut. All so called 'self-locking' nuts (see fig. 7.8) are better referred to as stiff nuts and (for structures) must not be used singly, that is, the minimum number of bolts and stiff nuts in a group must not be less than three. For some applications nuts are wire locked as shown in fig. 7.8. If bolts are used through fittings which are unlikely to be removed for checking during the service life of the aircraft then serious consideration should be given to the use of lighter fasteners of the hi-shear type (see fig. 9.14).

9.4.5

In addition to the solid rivets there are various types of rivets which can be placed without access to both sides of the structure. These are referred to under the general heading of blind rivets. An indication of the general principles is shown in fig. 9.14. The method of using almost all the blind rivets involves having a hole right down the length of the shank and with some types, such as the Chobert, this hole remains after the rivet is installed. In other types the hole is plugged automatically as part of the rivet setting process. All the various patterns have their own advantages and there is a great deal of commercial competition in this market. For discussion purposes, and since we have referred above to Chobert rivets, the companion self-plugging rivet from the same manufacturer is the Avdel MBC. Chobert are for light duties, comparatively cheap to buy, very cheap (in labour terms) to place and are produced in several different materials for different purposes. Avdel MBC are much more expensive to buy and to place than Chobert but are very strong (at least as strong as solid rivets) and apart from the facility of being able to place them from one side of a structure, the fact of mechanical installation means that vulnerable surfaces such as wing skins are not damaged by the hammering which is required to set up the strong types of solid rivet.

9.4.6

Wherever possible aircraft structures are designed so that bolts and rivets are used in shear. Some companies allow rivets to be used where they may be subjected to a small amount of tension but, in general, fasteners in tension must be bolts.

Fig. 9.15 and other illustrations in this book show the way fasteners are set out in aircraft structures.

Some particular points in fig. 9.15 should be noted. The distance A in fig. 9.15(b) must be sufficient to allow a riveting tool to have access unless a blind rivet is being installed from below. As a guide, for $\frac{1}{8}$ in diameter solid rivets in 0.04 in thick sheet dimension A should be at least 0.3 in and the edge distance should be 0.25 in. For blind rivets installed from below, A could be reduced to 0.25 in. The above dimensions should all be increased by 0.05 in for $\frac{5}{32}$ and $\frac{3}{16}$ in diameter rivets but these dimensions are sufficiently important that draughtsmen must check the standard practice in their particular company before quoting figures. Unsuitable edge distance allowance by the designer can cause quality control problems, i.e. a rivet hole positional tolerance of $\pm$ 0.02 in could result in a $6\frac{1}{2}\%$ strength loss in a 0.3 in edge and a 10% loss in a 0.2 in edge, but a 0.3 in edge is 50% heavier than a 0.2 in edge.

Some of these arguments apply to fig. 9.15(c). In this case dimension A must be determined by the clearance needed for wrench access. However, because of the load conditions indicated by the arrows, the edge distance must not be less than dimension A. Even if the two dimensions are equal, the tension load in the bolt will be twice the

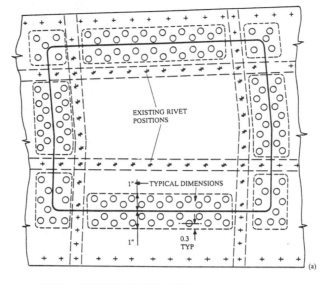

TYPICAL REPAIR TO LEADING EDGE

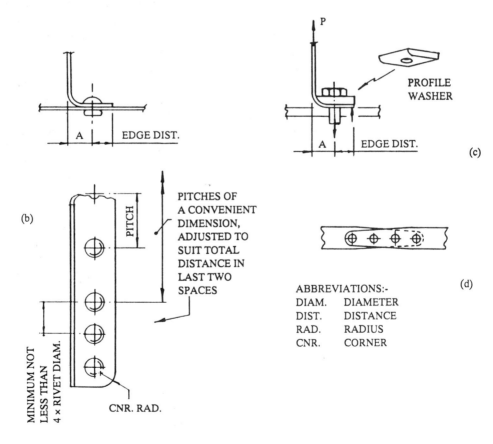

Fig. 9.15   Dimensions of riveted and bolted joints ((a) – *Courtesy of British Aerospace*)

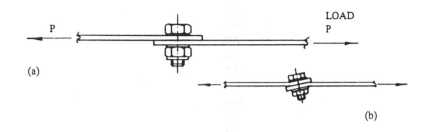

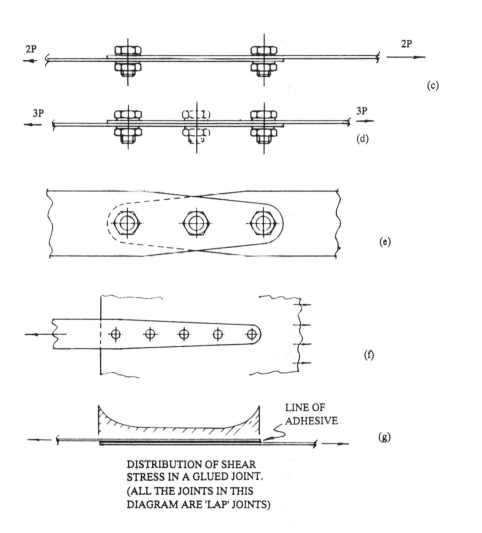

(a)

(b)

(c)

(d)

(e)

(f)

LINE OF
ADHESIVE

(g)

DISTRIBUTION OF SHEAR
STRESS IN A GLUED JOINT.
(ALL THE JOINTS IN THIS
DIAGRAM ARE 'LAP' JOINTS)

Fig. 9.16   Joints

load P in the angle. The use of the profile washer is necessary to avoid bending the flange of the angle, but its also helps keep A to a minimum.

9.4.7

The basic loading of bolts (or rivets) in a shear joint is shown in fig. 9.16(a). Ignoring any design criticism of the whole joint (such as the possibility of distortion putting the bolt in tension as shown in (b)) we can see that the shear load (see para. 5.5.4) in the bolt is P.

Similarly in fig. 9.16(c) we have doubled the load but shared it between two bolts so the shear in each bolt is again P. Now consider (d). In this case we have increased the load three times and increased the number of bolts to three, so that we are tempted to say that the load in each bolt is still P. Unfortunately, this is not necessarily true for the following reason:

If we consider (c), the material between the two bolts is carrying a load and is stressed. Because it is stressed it is strained (i.e. it is extended) and because the stress is constant over the whole length between the bolts the extension is even and uniform over the length. Now if we imagine (c) being kept in a loaded condition while a hole is drilled through both parts to receive the third bolt shown in (d) we can see (because the extension is even) that the holes in both parts will still line up and still accept the bolt after the load is removed. The situation is that the middle bolt can be placed in its hole when the joint is either loaded or unloaded, so it is quite clear that when the load is applied the middle bolt carries no load – that is, the two outer bolts in (d) each carry $1\frac{1}{2}P$.

In practice, because of minor distortions, the centre bolt does receive a load (but not a third of the total). Also we can taper the joint plates as shown in fig. 9.16(e) which has the effect of making the strain between the outer bolts uneven over the whole distance, and so by careful design we can make the loads on the bolts very nearly equal.

Although the example above is somewhat over-simplified, the underlying principle is that in a line of bolts such as fig. 9.16(f) the loads on the bolts are not equal. Also, in a joint where the plates are bonded together with an adhesive (such as epoxy resin), the shear stress in the adhesive is not constant along the joint but is concentrated at the ends.

Students who have become practised in consideration of stress and associated extension will be able to satisfy themselves that the shear distribution diagram shown in fig. 9.16(g) is representative. (It should be noticed that the diagram must be symmetrical because the joint is symmetrical.)

9.5   Joggling

9.5.1

Joggles are features of the type shown in fig. 9.4. In sheet metal structures they are virtually unavoidable. (Fig. 9.4(b) shows a method of avoidance but it is heavy and expensive and only justified where strength or tooling considerations demand drastic solutions.)

So far as the strength of joggles is concerned their problem is illustrated in fig. 9.4(c), but the difficulty is made more severe because, in spite of knowing that this failing exists, the designer can still be persuaded otherwise by the robust appearance of joggled components. If an attempt is made to react the unbalanced load P by hoping that a rivet head may take some tension (fig. 9.4(d)), and if the angle of the joggle is 45° then the load R is, by triangle of forces (see fig. 4.3), the same as P. Load R can be reduced by increasing the length of the joggle and hence reducing the angle in the triangle of forces, and this partial solution, carefully used, is satisfactory.

The illusion of strength in joggles is most marked when angles are joggled and it disguises the fact that the joint in fig. 9.4(e) is as poor as that in (f). A better solution is shown in (g).

9.6 Clips or Cleats

9.6.1

The two terms are used respectively in the American and British industries.

9.6.2

Typical cleats are shown in figs. 9.17(a) and (b). All sheet metal structures use these or similar connections and it must be remembered that they are structural components requiring the same stress analysis as more glamorous components.

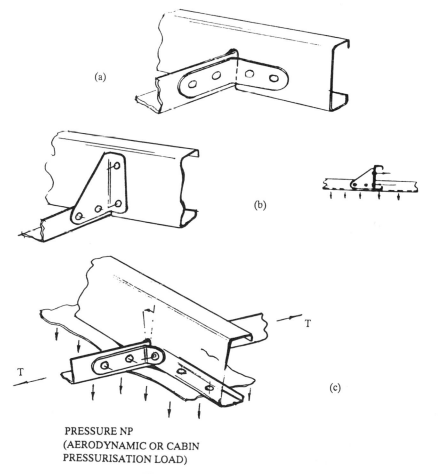

PRESSURE NP
(AERODYNAMIC OR CABIN
PRESSURISATION LOAD)

Fig. 9.17 Cleats or clips

9.6.3

Fig. 9.17(c) illustrates the fault which makes the use of single rivet or bolt connections highly suspect.

The reason for not being more definite and saying that such joints are 'impossible' is the slight reservation that if the only load involved was in the direction 'T' then the distortion shown might not occur.

These clips are sometimes called shear connections, and if this is true then rivets are adequate. If, however, the load is as shown in fig. 9.17(c) at *NP*, a typical pressure 'normal' to the skin (or perpendicular to the skin), then it must be remembered that, as shown in (b), the rivets can be in tension. In (a) the rivet tension is limited by the lack of torsional stiffness of the cleat between the bend and the first rivet.

It should be clear from the above remarks that the design of cleats is not a task to be undertaken lightly.

9.7   Stringer/Frame Intersections

9.7.1

A primary difficulty in the detail design of stressed skin structures made from sheet metal occurs at the point where the horizontal members (the stringers) cross the vertical members (the frames). Obviously at the point of crossover it is not possible for both members to be continuous. Some solutions to the problem are shown in fig. 9.3. Fig. 9.3(b) shows a particular arrangement from the Boeing 747, and we can see from the very careful shaping of the cut-out in the frame which allows the stringer to pass through, that the design has warranted, and been given, a great deal of thought. The cut-out is clearly an unavoidable notch in the frame, and the design effort has been directed to minimising the stress raising effects. The stringer clip (or cleat) and the joggling of the stringer have also been carefully designed to avoid the criticisms of paras. 9.5.1 and 9.6.3.

9.7.2

On an aircraft of the size of the Boeing 747 there are so many of these stringer/frame intersections that a company can justify an expensive design effort aided by computers and physical test programmes devoted to obtaining maximum efficiency from the engineering of this joint. Efficiency in this context means weight reduction without loss of structural integrity in the frame, and integrity means that the frame will perform all its specification functions without fatigue failure problems reducing the life of the structure.

All aircraft manufacturing companies have their own arrangements for this type of crossover joint, and students of design will need to follow the standards set within their company.

9.7.3

With machined structures the problem either does not arise or is not so pressing and just the avoidance of this one factor alone in the design would make the integral machined skin and stringer combination extremely attractive.

9.7.4

As structural components such as fuselages or wings are usually tapered, it is necessary to reduce the number of stringers towards the narrow end. Stringers should end at a frame and not in the middle of a panel. If they are stopped without being cleated (clipped) to a frame, movement between the very stiff frame and the stiff stringers, caused either by the flight loads or by vibration , will cause cracking in the relatively flexible skin. Although this particular point is a fairly obvious and easily avoided problem designers have to beware of similar problems in other areas such as galley structures, freight containers, etc., where a framework is covered with a skin which carries load.

9.8 Lugs

9.8.1

The present normal type of stressed skin structure does not use tension lugs of the type shown in fig. 9.18, but they are mentioned here because the ways in which they can break if overloaded are interesting and relevant to some related components such as the cleats shown in fig. 9.17.

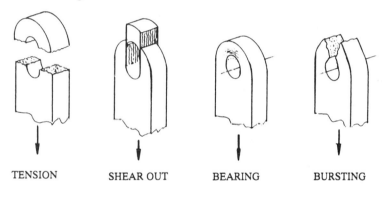

TENSION   SHEAR OUT   BEARING   BURSTING

WAYS (or MODES) OF
FAILURE OF LUGS

APPROX. PRO-
PORTIONS OF
A CORRECT LUG

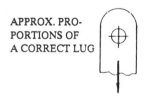

Fig. 9.18   Lugs

9.9 The 'Stiff Path'

9.9.1

We shall mention in chapter 11 (11.3.4) that some structures are unavoidably designed so that the applied loads can be carried in more than one way.

The example used is the floor panel which is attached on four sides but which would be quite capable of carrying sufficient load if it was only attached along two opposite sides.

In chapter 11 this is presented as an advantage for the stressman because he may find by analysing one *load path* and finding it strong enough on its own, that he does not need to go to the effort of finding the share of the load carried by each of the possible paths. However, in detail design the situation is slightly different. Here the designer must be careful to avoid the situation where he has (for instance) two load paths neither of which is strong enough to carry the applied load on its own (an apparently satisfactory design) but he also has one path more rigid than the other. He may then find that the flexible structure does not start to resist its share of the load until after the rigid structure (the *stiff path*) has exceeded its load-carrying capability and broken.

9.9.2

An easy trap to fall into is that set by the use of different materials in one structure. For instance, if a designer attempts to reinforce a light alloy structure with a steel addition he may find that the steel tries to carry all of the load, because the Young's Modulus of steel is much higher than that of light alloy; that is, the steel presents a stiffer load path. When calculating the stresses in a situation of this sort it is not only necessary to find the simple 'load over area' stress, but also to check the stress due to deflection (that is stress = Young's Modulus (E) $\times$ strain).

CHAPTER 10

Quality Control and Airworthiness

10.1   Quality Control

Quality control is complementary to aircraft structure and therefore only complementary to this book. However, the airframe engineer is so constantly involved with QC that it will be useful to give an indication of the methods used to ensure that the structure meets the specifications which it is intended to meet.

10.2   Quality

10.2.1

Everybody associated with an aircraft is concerned with its quality (especially those who fly in it) and a very large percentage of those associated bring a personal influence to bear. A cabin cleaner on a remote outstation and a radiographer examining a wing joint for cracks are both influencing quality.

10.2.2

In general industry outside aviation, product quality control starts at the top and is controlled by using the two-way management communication system. The board of directors of a company (or the company president) set the quality standard for their product according to their assessment of the market. Middle management and the engineering department specify the quality as part of the design and instruct the manufacturing staff on how to achieve the correct standard. The manufacturing staff then make the product to the standard required and offer it to middle management for acceptance. Middle management monitor the work and report back to the directors.

10.2.3

In the aircraft industry the process is somewhat modified by the influence of agencies outside the company. In the cases of both military and civil aircraft, these agencies set standards for design and manufacture which must be adhered to. In the civil aircraft market minimum quality standards are set by the Airworthiness Authorities. Table 10.1 shows the titles of the authorities in various countries. These authorities are backed by the law of their respective countries and publish very detailed requirements of design quality which must be achieved before the aircraft is permitted to fly. They also require the maintenance of quality on production aircraft to be kept at a continuing level of excellence which they endorse and monitor.

TABLE 10.1 Airworthiness Authorities

United Kingdom, CAA	Civil Aviation Authority Aviation House Gatwick Airport (South) GATWICK W. Sussex RH6 0YR Tel. 0293-567171 Fax. 0293-573999
Italy, RAI	Registro Aeronautico Italiano Via Del Trione, 169 1-00187 ROME Italy Tel. (6) 678 0951
France, DGAC	Direction Générale de l'Aviation Civile 93 Boulevard du Montparnasse F-75382 PARIS France Tel. (1) 454 43839
Germany, LBA	Luftfahrt-Bundesamt-Abt Flughafen 3300 BRAUNSCHWEIG Postfach 3740 West Germany Tel. (531) 39021 Fax. (531) 390 2254
USA, FAA	Federal Aviation Administration Chief Aircraft Engineering Division 800 Independence Avenue, S.W. WASHINGTON, D.C. 20591 USA Tel. (202) 366-4000

10.2.4

Because of the status of the Airworthiness Authority it would not be possible for a company making or operating aircraft to set low standards of quality for commercial advantage. In fact, since the earliest days of flying there has been remarkably little conflict between the Airworthiness Authorities and commercial interest, and most aircraft manufacturers set their quality standards above the already high legal minimum, as do the airline operators.

10.3 Control

10.3.1

Quality control starts with *education* to ensure that the newest recruit to the industry is aware of the importance of quality to the safe and efficient operation of aircraft. The second aspect is *checking*. Everyone checks his own work until he is satisfied with its correctness. Then the inspection department acts as a screen or filter to pick out the inevitable errors which we all make. Finally, the technical director (or chief engineer) and the Airworthiness Authority monitor the inspection department to ensure that necessary systems and procedures do not fall into disuse. The third aspect is the setting up of *systems* designed to make quality control more foolproof.

Typical of the many systems required for quality control are:

a system for issuing and distributing up to date corrections of engineering drawings;

a procedure for notifying the drawing office of errors found on drawings;

a system for ensuring that workshop measuring instruments are regularly checked for accuracy;

systems to ensure that parts of an aircraft found to be defective on one flight are repaired before the next.

The final aspect is *specification* (usually by engineering drawing or maintenance manual) linking the customer's and the Airworthiness Authority's quality requirements with the company's manufacturing or maintenance capabilities.

10.3.2

A major function of the Airworthiness Authorities particularly relevant to this book is the approval of the design of the aircraft and its equipment down to the smallest detail. The methods employed to examine the design and assess its quality vary in detail from country to country, but in general they are based on a system of acceptable reporting. This means that the authority recognises the competence of a company (or in some cases an individual), and when that company reports that a design meets the authority's requirements then the design is accepted as being 'airworthy'. This system operates within specified limits for each company; that is a company approved to issue reports on the airworthiness of engines would not expect to be considered competent to report on airframe structure. Also, the 'approved design organisation' or the 'designated engineering representative', is subject to constant monitoring in that every report is examined by the Airworthiness Authority and the relevant design only accepted when they are satisfied that their conditions have been met. The standards which a company must achieve before they can be considered as competent to report to the authorities are noted in each country's Airworthiness Requirements.

10.4 Procedures and Systems

10.4.1

Two examples are given to illustrate the working of quality control; both relate to aircraft structures in the manufacturing stage.

10.4.2

The first example concerns the procedure for dealing with defects in manufactured components found by the inspector. Because of the outside agency's influence on the industry, aircraft inspection is rather different from that in, say, the automobile industry. The aircraft inspector is employed and paid by the company he works with, but in the event of a conflict between commercial interest and quality, then his loyalty lies not with his employer but with the Airworthiness Authority. This sounds dramatic but in the event of a major snag or defect being found no company would give priority to any consideration of the commercial problems created by grounding an aircraft or scrapping a component. A minor defect, on the other hand is dealt with by a procedure as follows:

If the inspector finds that some items in a production batch of bolts (for example) have a thickness of head which is less than is allowed by the tolerance on the drawings, then he raises a *reject note* which he sends to the production supervisor with the faulty parts. (Tolerance is the sum of the variation above and below the nominal dimension. Nominal dimension and upper and lower variations (called limits) are shown on the drawing.)

However, he does not think that head thickness is a feature of such importance that the bolt should be scrapped. He therefore arranges with the production supervisor for *concession procedure* to be started. Initially a *concession note* is addressed to the technical department (or to the chief designer) requesting that the concession be made. The note, probably on a printed form, will identify the bolt by description and part number, say how many items are involved and probably include a sketch to clarify the defect. In their turn the various sections of the technical department (drawing, stress, weights) would examine the request and comment on the effect of the fault on interchangeability (for spares), strength or any other aspect within their interest. Provided that there is no adverse comment the chief engineer (or chief designer) signs the note to indicate his approval of the request which is returned to the inspector as an authority to accept the bolts as usable parts. This effectively cancels the reject note and the two documents (reject note and concession note) are filed together as part of the history of that order for bolts.

A similar document to the concession note is the *production permit*. This is used before the event rather than after. It is probably on the same printed form and follows the same route. A typical use is to request permission to produce a part in an alternative, more readily available, material.

10.4.3

The second example traces the route of a quantity of aluminium alloy from manufacture to conversion into a finished component.

(i) The alloy is cast and is given a *cast number*.

(ii) The cast billet is converted into extrusion and the extruded bars are marked with the cast number and a *batch number*. Test pieces are taken from the batch and checked to ensure that the material meets its specification.

(iii) The extrusion is sold to a stockist and is accompanied by a document signed by the manufacturer and certifying that the material meets its specification. This document called an *Approved Certificate* (or a *release note* or a *statement of conformance*) also notes the batch number and test report number.

(iv) The stockist gives the extrusion his own batch number and holds it in a store which has been approved by the Airworthiness Authority until he sells the material to various manufacturers.

(v) Part of the original batch is delivered to a company who receive several feet of extrusion and a delivery advice note. The 'goods inwards' inspector of the receiving company immediately places the material into quarantine store. Material in quarantine cannot be used until the inspector authorises its movement into bonded store, and he will not do this until he receives the stockist's Approved Certificate.

(Note that the original material manufacturer has issued one Certificate but the stockist may issue several relating to the original batch of material as it disperses to various users.)

The stockist's Approved Certificate passes on the assurance that the material is to specification and notes the batch number. The inspector again rebatches the material into his company's own system and notes the details (i.e. material specification, dimensions, supplier, date received, etc.) into his batch book. The material passes into stock and awaits further conversion.

(vii) Eventually part of the stock is withdrawn from store for conversion into a structural component. At this point it acquires its last (or probably its last) batch number. This number appears on a route card (or job card or work sheet) which accompanies the material from process to process until it is converted into a finished component. The component is finally stamped or marked (according to drawing instructions) with the last batch number and goes into service.

This laborious but simple and straightforward process which applies to all sorts of aircraft materials and components is a very old-established cornerstone of British aircraft quality control procedure. It has two fundamental uses. Firstly, anyone receiving any piece of material or equipment together with an Approved Certificate can use that material with confidence that it will perform its function according to its specification. Parts for which an Approved Certificate cannot be obtained can be considered as bogus parts of doubtful manufacture. Secondly, in the event of a component proving defective in service others with the same batch number can be traced. In the extreme case by tracing backwards through the various companies' batch books to the original cast number and then forward again all the components which were eventually made from the cast can be traced.

In the USA the system is different. Manufacturers of equipment and suppliers of material provide an affidavit sworn before a Notary Public that the material in question does meet the requirements and specification that it claims to meet. Under this requirement the major companies have the burden of assessing the minor or subcontracting suppliers, whereas in the United Kingdom the government, through the CAA, assists by maintaining a monitoring system on a select group of approved companies whose QC meets a recognised standard.

10.5 Further Notes on Quality Control Functions

Some of the functions of quality control and some of the terms and expressions used tend to be loosely applied, especially in smaller companies, and warrant some further comment. For example:

(i) A *statement of conformance* (or a letter of conformity) from a supplier referring to a piece of material is useless unless it states clearly to what specification or standard the material conforms. It is equally useless if the standard is unacceptable to the customer.

(ii) A *concession note* starts life as a request for somebody to change his requirement or lower his standards and therefore must be addressed to that person. Typically in (say) a subcontracting situation the note would initially be from the chief inspector of the subcontractor addressed to the chief inspector of the prime contractor.

(iii) *Batch numbers* can easily be allocated to any groups of material or parts received into a company but if a nut from batch A is fitted to a bolt from batch B the assembly must be given its own number or the chain of the system (para. 10.4.3) is broken. Finished parts should be marked with their batch number but as a quick examination of any random selection of equipment will show this is not always strictly enforced. However, if a complete piece of equipment is given a *serial number*, the manufacturer's records, i.e. route cards, history cards or similar systems, will show all the components which went to make that particular article and their batch numbers. Companies purchasing and receiving equipment for use on aircraft should ensure through their *goods inwards inspection* that these procedures have been adhered to.

(iv) The function of the inspector is that of a filter. If a faulty component eludes him he has made a mistake in his job; but the faulty component was nevertheless made by the production department and it is not possible for them to wash their hands of the responsibility.

(v) Material or parts cannot be withdrawn directly from a quarantine store. By definition a material in quarantine has not yet been proved to be faultless. If it has already been proved to be faultless it should have been moved (by authority of the goods inwards inspector) into the bonded store.

(vi) The example set out in para. 10.4.3 above illustrates an example of *traceability*. This synthetic word tries to sum up the features and systems which must exist and surround each component concerned with the aircraft. It should for instance be possible to pick any part and by looking at the numbers printed on it, trace back to the raw materials and also to trace the applicable manufacturing drawing, applicable stress calculations and test reports, and eventually an authoritative signture by which the designer acknowledges his opinion that the part will perform properly. The object and benefit is mainly to ensure that if something does not perform as it should, the reason can be tracked down and corrected.

For design/drawing offices, a QA system which ensures traceability of drawings and other technical documents is enough to meet the requirements of civil airworthiness authorities. However, purchasers of equipment, especially military customers, will also be concerned with *repeatability*, that is they will expect that the drawings etc. are sufficiently detailed to ensure that parts which are supposed to be identical to one another really are so, even if they are made at different times by different people. Conversely, if the component or feature in question is clearly 'one off', as might be the case with a specific repair or the mounting of a special piece of equipment for a single operation, then a good drawing office system should be elastic enough to produce minimal but sufficient paperwork, without having to apply all the effort it would put into documenting a production order.

10.6 Airworthiness Engineering

With the gradual increase and codification of the laws and rules which control aircraft design and operation, a need has emerged for an engineer who can take an overview of the whole design with the object of ensuring that the published requirements are met. In a sense, the work is the opposite of innovative design in much the same way that careful accountancy is the opposite of entrepreneurial enthusiasm. The suggestion is that the airworthiness engineer should not attempt to control design, but make it his job to ensure that designers (and operators) are aware of the confines within which they must work, also bearing in mind that the efforts of one specialist designer may unknowingly be affecting some other specialist's area.

This book is about structures, and the requirements governing their strength have been very stable over a long period. This condition produces exactly the atmosphere in which newly published legislation may be unnoticed by the designers who can then waste expensive time through working to outdated ground rules. The timely advice of an airworthiness engineer who continually keeps himself aware of revisions to the rules governing safety would prevent wasted effort.

Because of the element of cross-specialisation involved, the airworthiness engineer must have a wide general experience of the aircraft industry (although no doubt in time and by natural growth the airworthiness engineer will become the airworthiness department which will have people specialising in specific areas). He must also have a sufficiently open mind to accept that large sections of his carefully acquired knowledge may have to be drastically changed by some new legislation. More difficult to accept, he may also find that standards he has worked to for many years are not acceptable standards in another country.

The quantity of written material published by the authorities which is required reading for the airworthiness engineer, is so great that he cannot be expected to have every facet of the requirements at his fingertips. What he must know is where to start looking for information. For civil aircraft the basic source of information is contained in the Joint Airworthiness Requirements (JAR) which refer directly to the Federal Aviation Regulations (FAR). The British Airworthiness Requirements (BCAR) have been superceded but may aircraft which were originally designed to meet BCAR are still flying and when discussing their airworthiness, reference must be made to the older standards. Also, BCAR contained a lot of airworthiness engineering knowledge presented in a concise and readable format. JAR are subscribed to by almost all European countries with national minor differences to particular rules recognised as margin notes.

For military aircraft, the source document for UK requirements is DEF STAN 00-970. US requirements are contained in MIL Specs.

Other publications that an airworthiness engineer must be aware of are listed below, with comments on the areas of particular interest to structures engineers. (Note here that, in common with many sources of information, no-one should attempt to learn word by word what is written down. Examination assessors and quiz masters still seem to put some store by a person's ability to recite facts, but it is far better if you can say that the information you require is probably in book A, or if not there then in book B.)

Accident Reports, from a variety of sources, help by highlighting errors which may have been avoided by better engineering.

Occurrence Reports (MOR) by which people involved in the operation of aircraft bring to the attention of the airworthiness engineer any incidents which may have a detrimental effect on safety.

Civil Aircraft Inspection Procedures (CAIP) now superceded by the rather less useful CAP 562, *Civil Aircraft Airworthiness Information & Procedures* are a mine of information on aircraft workshop practices.

Repair Manuals are published by the aircraft manufacturer and either describe specific repairs for specific damage or set out general rules for dealing with damage met in service.

An American publication, AC43.13-1A, *Acceptable Methods, Techniques & Practices*, is another immensely useful and practical source of generalised information that the airworthiness engineer should be aware of.

Air Navigation Orders (ANO) are pubished by the government (in the UK) to specify minimum standards for the operation of aircraft.

Airworthiness Notices, by which the CAA circularise safety information which is not conveniently covered elsewhere. Subjects from the list in 1991 include AN50 *Deterioration of wooden aircraft structures* and AN64 *Minimum space for seated passengers*.

Service Bulletins and Alert Service Bulletins are issued by manufacturers to operators to inform them of problems which require attention, or of difficulties which have been met by other operators.

10.7   Maintenance Schedule

The above notes mainly concern manufacturing and design quality, but QC has to continue into the operating life of an aircraft structure. This is achieved through an approved maintenance schedule which operates a system of *checks*. For large aircraft these are listed and defined by the aircraft manufacturers, but to give an indication of how a typical schedule works the following would be reasonable for a light aircraft:

Check A — a pre-flight check at least before the first flight of the day
Check 1 — 50 flying hours or 2 months whichever is the least elapsed time
Check 2 — 100 flying hours
Check 3 — once a year or 500 hours
Check 4 — once every two years or 1000 hours

The following structure items would be among the details inspected at the various checks:

Check A — skins, moving control surfaces, windscreens and windows, door locks, all visually inspected for obvious defects
Check 1 — all fixed and moving structures, doors, windows, access panels, fuel tanks – inspect for visible corrosion, operation and damage
Check 2 — internal structures especially near potential moisture accumulation points – inspect for corrosion
Check 3 — structural joints – inspect for slackness
Check 4 — all structural parts – inspect for corrosion, slackness, cracking wear or any form of deterioration

Items found to be defective at any inspection would be rectified before the aircraft continued in service.

To avoid taking aircraft out of service too frequently or for long periods it is often the practice to agree a system of running checks with the CAA. Under such a system all the items required to be checked yearly (say) would be listed, and instead of all being checked at one grounding of the aircraft they would be attended to in weekly or monthly batches.

10.8   References

Although it is the responsibility of companies to set up their own quality control systems, advice and help is available from the publications of the Airworthiness Authorities and inspection procedures in particular are well documented. Lists of CAA publications are available on request from:

Civil Aviation Authority
Printing and Publication Services
Greville House 37 Gratton Road
Cheltenham   Glos. GL50 2BN
Telephone 0242 35151

AC 43.13 – 1A & – 2A
is obtainable from:
IAP, Inc.
P.O. Box 10 000
Casper
Wyoming   82602-1000

CHAPTER 11

Stressing

11.1 Introduction

11.1.1

Stressing, or stress analysis, is primarily that process which estimates whether or not the proposed structure is strong enough to carry the loads which will be imposed on it by the operation of the aircraft.

The designer (or, with the complexity of modern aircraft, the design team) is required to design a structure which will allow the aircraft to perform in its role efficiently. There are many constraints which influence the design such as maintenance accessibility, production costs, aesthetic appeal and so on, but this chapter is only concerned with the structure and its problems.

The overall loads applied to the structure are determined by domestic airworthiness requirements together with those of the countries in which the aircraft will be sold. These required loads, the specification of the role of the aircraft and the nature of the constructional materials and manufacturing processes available, are the raw data with which the designer begins.

It is in the nature of aircraft that the lighter they are the better they perform in their role. Also, commercially, the cost of carrying superfluous weight in terms of total fuel consumed during the life of the aircraft is very high. The designer's task then includes achieving maximum lightness and to do this he will employ the most advanced materials and constructional techniques which are within the production capability of his company. (The last sentence is not intended to suggest that complexity is at any time desirable. Simple solutions are almost always the best for any design problem. One of the great aircraft designers, Ed Heinemann, said 'Simplicate and add lightness'.)

Going hand in hand with the advance of materials and techniques is more and more detailed and accurate analysis by the stressman. This has a twofold aim: to reduce weight by ensuring that every piece of structure is fully employed and also to eliminate expensive test failures. Airworthiness approval only occasionally relies on the stressman's calculations; there is always sufficient testing on major structural items to confirm compliance with requirements, though tests do not necessarily show up components which are over strength. Also, with larger aircraft and more expensive manufacture, companies are very keen to eliminate any early failures on test. In spite of the fail safe type of design philosophy, premature failure of one component can scrap a whole test piece structure.

So, if the primary stressing task is to ensure adequate strength, the secondary task is to assist the designer with the aim of producing structures which are exactly strong enough and no more.

11.1.2

Stressing is now a highly complex subject relying heavily on mathematical analysis. In this chapter the techniques used are discussed in their most elementary form, with the idea, as stated in chapter 1, of giving a foundation vocabulary on which readers can build a more detailed study.

Although stressing is such a mathematically biased subject it is not necessary for stressmen to be more than competent mathematicians. The majority of the work done in the examination of structure relies on the use of data and formulae which only require practice and sensible application for their satisfactory use. Writing the data and evolving the formulae do require considerable mathematical skill but the stress office always has a place for the type of engineer who has a 'feel' for structures and who can recognise the weak points in two minutes, as well as the man with several degrees who can identify the same weaknesses after two weeks of calculating. Both types of stressman are useful and even essential to a balanced team.

11.2 The Stressman's Work

11.2.1

When a stressman examines a piece of structure he must

(a) determine the load distribution throughout the structure
(b) determine the extent to which the elements of the structure are capable of supporting the loads imposed on them.

These statements are explained in the next section.

11.2.2

Statement (a) is simple enough when we consider a simple structure surrounded as we said in chapters 4 and 5 by a balanced system of loads and reactions. If a thin chain is suspended from a hook in the ceiling and a weight put at the bottom of the chain, then the weight is the applied load, and what happens at the hook is the reaction. At every link in the chain we have the same situation: an applied load at the bottom and a reaction at the top. (If we ignore any weight in the chain itself, the loads and reactions

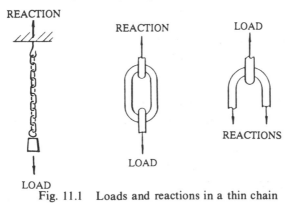

Fig. 11.1   Loads and reactions in a thin chain

are numerically the same all the way down.) If we now look at half a link we realise that what we were calling a reaction is in fact a load when we think of the link in isolation, and the reactions are provided by the other half of the link. (One bit's load is another bit's reaction.) By progressively thinking our way through this simple structure we have determined the load distribution.

Many structures can be examined and the internal distribution of loads analysed by statics alone, that is, by considering in careful stages how each load is reacted and by progressively dividing the structure into sections (as we did with the chain) balancing loads and reactions as we go. Fig. 4.3 earlier in this book is relevant here.

Unfortunately aircraft structures usually fall within a group known as *statically indeterminate structures*. These are structures which have more than one load path and fig. 11.2 shows two classic examples. Frame (a) is a structure which is sound enough but it is also still a satisfactory structure at (b) with one member removed. Beam (d) is a good structure but it would still hold up the two loads if it was divided at

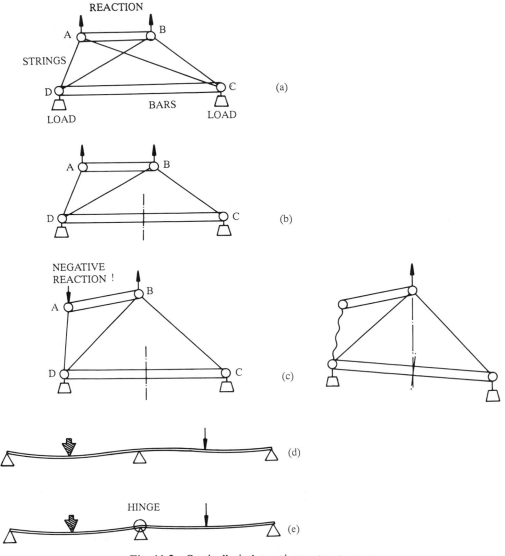

Fig. 11.2   Statically indeterminate structures etc.

the middle support. The continuity at the centre support in (d) and the removable member in (a) are called redundancies (or redundant members) and the load distributions in statically indeterminate or *redundant* structures have to be found by consideration of the deformations.

(We can use fig. 11.2 for consideration of two other ideas. If we remove the other diagonal (BD) from (b) the structure would cease to be a structure and would become a mechanism. That is the bar (CD) would swing into some position of repose which would not be where we wanted it to be. The other thought is that members (AD) and (BC) are in tension. It happens that the load in (BC) is greater than the load in (AD) and also (BC) is longer. If we are not careful with their sizes the two members could stretch as shown in (c) so that the line of action of the centroid (i.e. the centre of gravity, CG) of the two weights could come outside point B which would put member (AD) into compression. If (AD) was not capable of carrying compression the structure would have changed into a mechanism and what appeared at first to be satisfactory would have turned out to be quite the reverse.)

Determining load distribution can be difficult but with practice and thought most problems can be solved. Chapter 5 emphasised the importance of the relationship between stress and strain. the vital key to a clear analysis of some very complex structures is the very simple idea that there is no stress without strain and, equally, if a part of a structure is strained it is also stressed. This concept is applied in an interesting way in the consideration of the effects of kinetic heating (see para. 4.4.3). Because these effects and ideas are not easily understood we will digress to look at the example of a simple structure which is not an aircraft part, but will illustrate principles that do apply to aircraft structures.

Imagine a camera tripod with its feet firmly anchored to the ground (set in concrete). Now imagine a fourth leg added to the tripod. Provided that the extra leg is exactly the correct length, its influence is zero, but the structure which was three legged and needed all its members is now redundant, or statically indeterminate; that is, any one of the four legs could be sawn through without the camera falling. Obviously, we cannot affect our camera support by kinetic heating, but we could heat one leg with a blow lamp. If this one structural element is heated it will expand, or try to expand, by an amount which can be calculated by multiplying the temperature rise by the coefficient of expansion (a known physical property of the material). Because of the heating, this extension takes place without any stress being produced, but if the extension is prevented or reduced by the reactive efforts of the other members (as it will be in this case, since they are anchored to the ground), then the amount by which the extension is forcibly reduced is effectively a compressive strain with consequent stress and hence load in the member. We now have the heated member in compression, being loaded by the other members, which in our example, are in tension; all these loads have been produced by heating just part of the whole structure. Structures of this type with more members than they need, but with a different form (stressed skin) are normal in aircraft and are affected by kinetic heating in a similar way to the way in which the camera mount was affected by the blow lamp.

The situation of one member's load pushing or pulling on another member is typical of all statically indeterminate structures under any externally applied load (apart from or in addition to the heating condition). Usually in aircraft structures the problem is complicated because as well as straightforward changes to the lengths of the members, there are also angular deflections and rotations at the joints between members to be considered. However, the principles are simple and if we compare, step by step, the extensions, the strains, the stresses, and consequently, the loads, we can arrive at the condition where all the loads are balanced, with one member's load reacting the next member's load, and the problem is solved.

11.2.3

Once the distribution of loads around the structure is quite clear (and do not forget that, as we pointed out in para. 4.2, there is more than one case to be considered), each element of the structure has to be examined to see whether or not it is capable of carrying the load applied to it. Apart from the very common possibility that the stress level is too high, the stressman must be sure that members are not fundamentally incapable of carrying the type of load imposed. A typical example of this sort of problem is one that frequently causes dissent between draughtsmen and stressmen. The draughtsman will fasten two components together with rivets (which are good shear connectors but are incapable of taking tension) and the stressman will say that they must be bolts. If at this point the draughtsman shows good reasons for not having bolts then there will inevitably be disagreement.

11.2.4

Arguments of the type given as an example in the last paragraph can be avoided if the draughtsman and the stressman work together from the beginning of a design task. If such a partnership can be made to operate smoothly it can be of great benefit to the job. Inevitably it is not an easy relationship since the draughtsman is inclined to think of the stressman as a head-in-the-clouds academic while the stressman is certain that he could make a better job of the design if only he had more time. Neither of them is right. The designer/draughtsman is bound to know more about manufacturing processes and design aesthetics than the stressman because that is the field in which he has gained his experience. On the other hand, if he listens to the stressman's advice on load distribution and size of members, he can possibly pre-empt the inevitable day of reckoning when all structure design work still on the drawing-board is declared by the chief designer to be both too weak and too heavy!

11.2.5

The remaining part of the stressman's work is to continue his academic studies, either formally at evening school, or informally through reading and discussion with others in the profession.

11.3 Stressing Methods

11.3.1

There are no short cuts to learning stress analysis. Study and work under the instruction of experienced stressmen are the only routes. Some approaches to study are, however, easier than others and design draughtsmen or licensed engineers who would like the satisfaction of designing a repair scheme, or even a piece of structure which the stressmen can approve without argument, should be able to teach themselves enough theory by reading.

Some knowledge of mathematics is essential and some further knowledge gives added interest. The essential knowledge is basic algebra (for the manipulation of formulae) and trigonometry (for the solution of triangles). A knowledge of calculus and how to read the mathematical shorthand which goes with it opens a whole field of interest, because without this it is almost impossible to read the textbooks and understand how the formulae are evolved.

Schools and colleges sometimes fail to explain the connection between pure mathematics and the study of practical subjects such as the theory of structures. An ability to use maths with ease, enables the engineer to set up mathematical models. These are close approximations to the true and accurate situation, which can be thoroughly analysed by mathematical methods. A typical example in the stressing of beams would be a beam with a loading pattern so close to being evenly distributed that the stressman could fairly assume that it was exactly uniform. This assumption of exactness means that the next stage of calculation examines a truly parabolic bending moment distribution and the parabola is one of many geometric shapes which can be integrated or differentiated or otherwise mathematically dissected with ease by someone familiar with the processes.

Extra practice with an increasing knowledge of mathematics enables the analyst to set up 'models' which more closely and accurately represent more complex problems.

The remainder of this chapter notes some things which are not usually mentioned in textbooks and also continues the policy of explaining some of the terms and expressions used in stressing.

11.3.2

The best way to start stressing a piece of structure (perhaps the only way) is to stand back and look at it. At the stressing stage it will only be a drawing so 'looking at it' means understanding the drawing and visualising it in your mind. It is a piece of structure so it is being loaded. (If there is no load we are not looking at a piece of structure but at a fairing or a piece of in-fill.)

As it is being loaded the loads are being reacted, so we look for the places where the loads are coming in and where the reactions are being applied.

The structure is a link between the loads and the reactions and between the two it will be pushed or pulled or sheared or bent or twisted, or it may be subjected to all of these, some of them simultaneously. We must think about it and be absolutely sure that the structure is arranged to cater for all the loads being applied to it. It is ridiculously easy for a designer to cope admirably with a load in one direction and entirely forget another load in another direction.

11.3.3

Once we are satisfied that the structure is complete we can begin the arithmetic. The first move in the calculation is to balance the external loads, that is, to make sure that every load is properly reacted. Checking the balance of the external loads is a continuous process right through the analysis of the structure. Every component must be in balance in every loading case. Being in balance means that the sum of all the loads (or the forces) and all the moments is zero. To check this easily, forces are divided (see fig. 4.3) into parts along three axes which are mutually at right angles. The usual axes taken are the x axis fore and aft along the aircraft, the y axis left and right across the aircraft and the z axis up and down. One of the most difficult things at this stage is keeping track of whether the load is positive or negative, but if we keep checking for zero balance any sign error will soon become apparent.

11.3.4

At this point the going is almost certainly becoming harder. Being an aircraft structure, the piece being examined stands a very good chance of being statically indeterminate, and as mentioned in para. 11.2.2 such structures need special methods of analysis.

With the help of computers it is becoming quite easy to use the finite element method. The practice is to divide the complex structure into elements or parts which can be examined separately, and their deflections under load matched with their neighbour's deflections. The structure of fig. 11.2(d) can be treated as two elements and solved without the help of even a hand-held calculator. However, structure 11.2(a), which is still not very complex, has six elements and if we had chosen to put joint plates at all the intersections instead of having pin joints (i.e. joints with clevis pins) the analysis would have already become too tedious for hand calculations.

In recent years powerful computer programs have been written to deal with this sort of problem and are now accessible to most stressmen, so that after a short study course on how to divide the structure and how to present information to the computer, structures which a few years ago could only have been analysed approximately, can now be examined very accurately.

Although this approach is still in the realms of the large company and the advanced structure, even humble equipment manufacturers will very soon be using these methods to refine the structures of their products. There are already several very usable, cheap programs for desk top computers, which are readily available and capable of analysing simple redundant structures.

Approximate solutions are often sufficiently accurate for smaller or ancillary structures, especially where the weight penalty of excess structure is being carried for other reasons. (A typical example here is a piece of floor being analysed for a load in the middle of the panel. Being held all round its edge the structure is statically indeterminate, but it might prove strong enough if only two edges were held.) A stressman faced with alternative load paths might quite convincingly and reasonably imagine all the load going on one path and if he found that strong enough he might not look further, satisfied that if one path was strong enough, the two paths would be stronger. Finding and applying approximations to load distribution problems requires the knowledge which comes with experience and discussion, but the majority of senior stress engineers are always ready to talk over problems and pass on a little 'know how'.

11.3.5

The last paragraph glossed over the problems of load path analysis. These can be very difficult and it is a part of the stressman's work which requires more imagination and insight than other parts which can be operated with data sheets. Having made that admission, it is also true that licensed engineers and draughtsmen who attempt design stressing soon develop a pattern of thinking which proposes simpler structures with clearer load paths to their own advantage and to the benefit of the aircraft.

11.3.6

Once the loads in the members of the structure are determined, calculating the actual stress level is usually straightforward, given that an adequate library of data sheets is available. One of the books listed in para. 11.5 (*Handbook of Aeronautics* No. 1) was written by the compilers of many of the initial Royal Aeronautical Society data sheets and it reproduces all those which are of fundamental importance.

The talent that the stressman brings to the job at this point is the ability to pick the correct data sheet (or the correct formula). To do this he must visualise how the member or component he is considering will fail. (For readers who may be fainthearted potential air travellers, the phrase '. . . will fail . . .' is another piece of jargon meaning '. . . how the component would eventually break or collapse if the application of load was to be continued past the required level'.) Once he has reached a decision on the mode of failure (i.e. the way it is going to break) he refers to the appropriate data sheet which will present him with one of the following:

(a) a permissible load on the component (e.g. the permissible shear on a specific rivet and sheet metal combination);
(b) a stress level which he must match against the specified capabilities of the material used;
(c) a factor by which a calculated stress or load must be corrected when the original calculation was not sufficiently refined. For example a rectangular section bar which has a hole drilled through it is loaded in tension. The apparent tensile stress is the load divided by that area of section of bar which is left either side of the hole. We know, however, from previous reading (chapter 9) that the hole is likely to be a stress raiser. So we look for information on such a condition (the author found it in *Formulas for Stress and Strain*, Roark and Young (para. 11.5)) and find that the first calculation of stress will need to be increased by a substantial amount (depending on the proportions of the bar the factor will be of the order of 2.0 to 3.0).

The factor does not alway indicate a stress increase. For instance some beams which appear to be overstressed on first investigation may be satisfactory after the application of a stress reducing factor relying on the proportions of depth to length, called a *short beam bending factor*.

The majority of stressing data is well presented and easy to understand. Often the compiler includes a worked example. For people who are not actually working in a stress office the most difficult type of data sheet to obtain will probably be that dealing with the strength of riveted joints. All the major aircraft manufacturers publish their own tests but only for their own people. A very short table of rivet strengths is shown in Appendix A at the end of this book.

11.3.7

In a large proportion of investigations, unless there has been very close co-operation between draughtsman and stressman from the beginning of the design, the first run through the stress calculations will reveal faults in the design. As these are corrected it can happen that the load distribution is affected and so the whole process must be re-run, but unless exceptional accuracy of design is being sought, two passes at the problem are usually enough.

11.4 Stress Reports

11.4.1

The object of the stress analysis is to determine whether or not a structure is strong enough and to achieve this the analyst compares the stress in each component of the structure with the allowable or permissible stress for the material of which the component is made. Sometimes as in the case of bolts and rivets the comparison is made between load and permissible load but the effect is the same. At the end of what may have been a massive calculation the stressman notes the actual stress level which he estimates will be in the material, and the stress report finishes on one line which says in essence: 'the permissible stress is greater than the actual stress, therefore the component is strong enough'.

There are two common ways of quantifying this. The British method usually quotes a *reserve factor*, and the American method quotes a *margin of safety*. It will be clearer to use figures to explain the difference. (The examples use imperial units, but followers of SI units will understand the points made.)

We assume that the stressman has investigated the specifications and the requirements, and has found the loads on the whole structure. He has multiplied by proof and ultimate factors, found the loads on the particular item in question, and calculated the stresses in the item. He then notes down (for example)

max. tensile stress $= 45\,530$ lbf/in²
permissible UTS for . . . material $= 51\,000$ lbf/in²

Therefore (in the British system)

$$\text{reserve factor (RF)} = \frac{\text{permissible}}{\text{actual}}$$

$$= \frac{51\,000}{45\,530}$$

$$= \underline{1.12}$$

or therefore (in the American system)

$$\text{margin of safety (MS)} = \frac{\text{permissible} - \text{actual}}{\text{actual}}$$

$$= \frac{51\,000 - 45\,530}{45\,530}$$

$$= \underline{0.12}$$

(Note that $MS = RF - 1$)

Reserve factors are usually noted in a right hand margin and the report (for one item) ends:

max. tensile stress $= 45\,530$ lbf/in²
permissible UTS $= 51\,000$ lbf/in²

$$RF = 1.12$$

(Note here that RF<1.0 – read 'a reserve factor less than one' – means that the strength of that component is inadequate and RF>1.0 – read 'RF greater than one' – means that the component is unnecessarily heavy.)

Note also that the RF (or MS) is assessed against the maximum stress found. In an earlier example (para. 11.3.6) we saw the effect of a 'stress raiser', but in fact that maximum high stress would be localised in a very small volume of the material. So if the RF is found for the place of maximum stress, as it has to be, then in our example, and also in most other components, there is a great volume of metal where the RF>1.0 and weight is being lost, which takes us right back to para. 6.2.1 where we discussed why steel is a heavy material to use, compared with aluminium alloy.

As aircraft design and manufacturing techniques improve we will progress towards the situation in which every ounce of material is fully worked and we have $RF = 1.0$ throughout the structure. At that point, the arguments in favour of using titanium may be overwhelming unless an even better material has been found.

11.4.2

The justification for discussing the last line of the stress report before the rest is that we need to see the ultimate objective.

(Incidentally, in any technical report it is a good idea to make the first paragraph, after a description of the objectives of the report, a summary of the conclusions. The rest of the report is then a justification of the conclusions and a detailed report of how the end result was reached.)

We said at the beginning of para. 11.4.1 that the stress in every component is investigated. This is certainly true for all the components which are newly designed for the structure being investigated, and a reserve factor (or margin of safety) is found and noted for each part identified by its unique drawing number. It is also generally true for load-carrying items of standard hardware such as bolts and rivets; the only type of exception being where, for instance, a group or line of rivets is so arranged that either the loading is evenly spread or it is clear that one rivet is more highly loaded than the rest.

11.4.3

So far we have the report listing reserve factors. It should also make a statement noting the minimum reserve factor found throughout the whole structure being reported. This would be part of a summary statement which effectively says that the stressman has examined the structure in question and finds it strong enough, within certain limits. Having made this statement, the report must also list those limits. These are the specifications and requirements referring to the aircraft in its role; any exceptions to specification or limitations of role, and a note of any assumptions made during the calculations leading up to the report. (Assumptions may concern conditions external to the structure about which no exact information was available at the time of the calculations, or they may concern approximations of the calculation method by the simplification or idealisation of a component which is difficult to analyse. The floor panel example in para. 11.3.4 would be analysed on the assumption that only two edges were held.)

11.4.4

The whole report will then list:

— Document reference number
— Issue or revision letter
— Date
— Title: Stress Report
— Subject: The structure identified by drawing number and name

(If it is a long report there may follow an index and a list of effective pages with their own current revision letter. Some reports, particularly those which present a summary of a number of minor reports, may need a revision record wherein it is noted that issued amendments have actually been incorporated in the text.)

— Summary of results and conclusions
— Associated specifications and requirements
— Cases: a list of cases considered
— Assumptions: a summary of major or significant assumptions made during the calculations (see para. 11.4.3)
— Exceptions and limitations: (to specifications)
— Calculations: the detailed calculations showing how the reserve factors are obtained. Apart from the arithmetic this section should justify the assumptions by giving the reasons for their use and it should show by reference where the data used in the calculations originated.
— Compiled by:
— Checked by:

The presentation of reports varies from company to company, and the easiest course before attempting to write one's first is to look and see how previous reports were set out. It is a personal view that typing is unnecessary and, because of the prevalence of mathematical formulae, not always successful. This does not apply to the *Type Record* which summarises all the reports for the whole aircraft. Alternatively, typing everything down to and including the 'Exceptions and Limitations' but leaving the actual calculations handwritten makes a neat job.

11.5   References

There are many books on stress analysis and the theory of structures and eventually everyone finds their favourite. In the following list of books, each contains its own bibliography so that readers are set off on a path of study which can be as detailed as they wish. The comments are my own but I think most stressmen would agree with them.

Bruhn, E. F., *The Analysis and Design of Flight Vehicle Structures*, Tri-State Offset.
— Very expensive in the UK but a master reference and guide. Full of easy to follow worked examples but entirely professional.

Roark, R. J. and Young, W. C., *Formulas for Stress and Strain*, McGraw-Hill paperback.
— A standard reference book stuffed full of practical and usable data. Early editions are closer to the aircraft industy than the 5th edition which gives wide coverage of all structural work.

Peery, D. J., *Aircraft Structures*, McGraw-Hill Book Co.
— A classic textbook with just the right blend of practical illustration and mathematical explanation.

Argyris, J. H. and Dunne, P. C., *Handbook of Aeronautics*, No. 1, *Structural Principles and Data*, Part 2, 'Structural analysis', London: Pitman.
— Mathematically very heavy but the maths lead to a mass of clearly presented and very usable data and formulae.

All the above are written using imperial units. For people used to SI units:

Megson., T. H. G., *Aircraft Structures for Engineering Students*. London: Edward Arnold.

CHAPTER 12

Presentation of
Modifications and Repairs

12.1 Definitions

12.1.1

As the title says, these notes concern modifications and repairs and by that we mean changes of greater or lesser degree to an existing aircraft. When an aircraft is first designed and built its features and characteristics are all recorded on drawings and reports, the structure in particular being documented in great detail. All these records are an integral part of the machinery of approval for the aircraft, approval which is given by the Airworthiness Authority (see para. 10.2.3) and which is required by law or by government order, before the aircraft can enter service. If now the existing aircraft is changed in any way, however small, it is no longer exactly the aircraft which was approved and the records can prove it. Therefore, when the change is made, if the aircraft is to continue to fly legally, the change needs approval from the Airworthiness Authority and documents of its own to supplement the original records of the aircraft type.

12.1.2

Changes to a design need some definition (and here we are dealing only with structure). Repairs clearly involve a change unless new identical spare structural components can be fitted using the original fixings and fixing holes. Most aircraft have a repairs manual which lists and illustrates standard repair schemes which are already approved and can be incorporated without further discussion or paperwork. Damage which is outside the range of the repairs manual will need a new repair scheme designed by an Approved Design Organisation (see para. 10.3.2) and accepted by the CAA. In most repair situations the Approved Design Organisation will be the original aircraft manufacturer.

   (Note: In this section we are describing British procedures. Constant reference to two different procedures is difficult but the systems discussed are fairly general. The American system uses consulting engineers (designated engineering representatives) to monitor design effort whereas the British system only accepts design effort by a designated (or approved) organisation.)

   Modifications are changes and under the CAA procedures can be either major or minor. For a definition of the difference we cannot do better than refer to British Civil Airworthiness Requirements Chap. A2-5.

'A modification will be classified as minor or major according to the nature and extent of the CAA investigation in connection with its approval. Where the investigation indicates that the particulars given in the Certificate of Airworthiness, or associated documents, will need amendment (even though no physical change to the aircraft is involved) the CAA may require major modification procedure to be followed where the amendments are significant.'

In para. 5.8 we defined Class 3 components; if the modification concerns a Class 3 part it can safely be classed as minor. Any other modification should be thought of as major until the CAA specifically agrees otherwise. Theoretically minor modifications can be designed by unapproved organisations. Modifications as they affect structure or structures designers are usually initiated by some operational requirement. They can range from enlarging a fuselage door, through designing new brackets and attachments for the latest navigation aid, to substituting a new type of rivet for one which is unobtainable. The first of these is obviously major but we cannot definitely say that either of the others is minor.

12.1.3

When an operator requires a modification on one of his aircraft, or on all the aircraft in a fleet, he approaches an Approved Design Organisation which may be inside his own company. If he needs to go outside his company he must be sure the office he approaches is competent and approved to work in the particular field in question. For instance, if the modification is structural, it is no good talking to experts in fuel systems. If the modification involves two disciplines (for instance, a new navigation aid may require major structural modifications to house an aerial in addition to electrical system changes), then one design office is picked to supervise the work and subcontract those parts which it is not qualified to handle. The approval of organisations is mentioned in chapter 10 with an indication of their necessary qualifications. In the context of this chapter, an organisation is recognised by the CAA as being competent to report to them that a proposed change to an aircraft will in no way produce a conflict with the British Civil Airworthiness Requirements. How the Approved Design Organisation justifies its confidence in its claim to itself and to the CAA is to some extent its own affair, but its methods and records will be open to scrutiny by the CAA as part of their assessment of competence.

12.2 The Paperwork

12.2.1

Every job starts with a specification and an order. In the case we are discussing the specification must come from the operator of the aircraft. It will probably be verbal and it may only say 'we have a problem here which must be solved within British Airworthiness Requirements', but hopefully it would be more complete. A well-written specification virtually designs the job. If the specification is verbal the design organisation would be prudent to write it down and feed it back to the operator for his approval (by signature). The order (so far as the technician is concerned) need only say 'Proceed with a design to my specification', but again this should be written and signed.

12.2.2

The next step is for the Approved Design Organisation to notify the authority that a modification has been proposed. The CAA use a form AD282 titled 'Application for Approval of Major Modification'; under the jurisdiction of other authorities a letter would be required. Whether by form or letter this action enables the authority to open a file and the modification procedure is set in motion.

The information required will probably be as follows:

Aircraft type
Nationality and registration marks
Constructor's number
Applicant's modification number
Name and address of applicant
Location of aircraft for inspection
Nature of modification
Original drawings affected
New drawings introduced
    and information on how it is proposed to effect any necessary corrections to flight
    and maintenance manuals.

12.2.3

While the first step is being taken the next steps will already have begun. The Approved Design Organisation will have started work on their design and drawings. These drawings have at least three functions; they communicate the design to the workshop and they record what the workshop did. The third function is to carry signatures by the stressman, the designer and any other signatories who the CAA agree are necessary to assure the quality of the design. So far as approval is concerned the record and the signatures are the important parts with the record being sufficiently clear so that, at some future date, all concerned could say with certainty how the British Civil Airworthiness Requirements were met. This means that the drawing must show at least all the dimensions affecting strength and any other applicable aspects of performance, all the material specifications used (by reference number) and details of standard hardware used.

12.2.4

As we noted above, the drawing will be signed by the stressman to indicate his approval of the strength of the change. Before he signs he must have justified his confidence to himself and he must also be in a position to justify his confidence to the CAA. To do this he must either make a satisfactory written analysis of the strength and write a stress report (i.e. he has got to stress it himself) or he must have available a satisfactory report of a physical test of the strength. In some instances he will have a strength test covering one particular loading case and the other cases will be covered by stressing. A test report will need to be preceded by a test specification set out with the following information:

(a) a reference number
(b) an issue or revision letter
(c) a date
(d) a title 'Test Specification'

Then the following headings and information:

  (e) Subject
  (f) Reference documents and relevant specifications
  (g) Object of test (usually to show compliance with specifications (f))
  (h) Method of test
  (j) Equipment needed
  (k) Loads to be applied
  (l) Method of recording load

A test report will need the following:

  (a) a reference number
  (b) an issue or revision letter
  (c) a date
  (d) a title 'Test Report'

Then the following headings and information:

  (e) Subject
  (f) Object of test
  (g) Date of test
  (h) Witnesses
  (j) Test specification reference number
  (k) Account of test
  (l) Results achieved
  (m) Conclusions (comparison with test specification)
  (n) Signature of tester and a witness

The stress report and its contents were discussed in chapter 11.

12.2.5

At this point different Airworthiness Authorities have various views on the document which gathers in the loose ends. The prudent design organisation might for its own record write an integrity report. Such a document would gather together, in the form of references to other paperwork, how representative the test piece was relative to the final installation on the aircraft, the origins of the materials of manufacture of the test piece relative to the materials to be used on the proper installation, the relationship of the test to the requirements and the relationship of the test results and the design to the ambitions expressed in the original specification.

The Airworthiness Authority would certainly require some form of declaration or statement that the modification or repair met with their requirements. The form of declaration asked for by the CAA is as follows:

(After listing the aircraft type, registration marks, C. of A. category, etc.)

I hereby certify that, except for the differences resulting from the modification(s) listed above, the design of the above aircraft has not been changed in any way. I further certify that, with the exceptions listed below, the design of the modified aircraft complies with the CAA Requirements so far as this particular type of aircraft is concerned.
EXCEPTIONS   Signed

(Then follows a list of deviations from requirements and a signature, being that of the person who represents the company to the CAA and who is known as an Approved Signatory.)

Other authorities ask for a 'statement of conformance' or a 'declaration of design and performance' all with the same commitment on the part of the design organisation.

12.2.6

The last piece of paper, the last word that justifies the structures engineer's efforts and recognises his expertise is the Airworthiness Authority's note of acceptance. In British terms this is the Airworthiness Approval Note (the AAN).

12.3   Reference Documents

British Civil Airworthiness Requirements
   (Sections A, D, G and K)
Civil Aviation Authority
P.O. Box 41
CHELTENHAM
Glouc.

Federal Aviation Regulations
   (Parts 23, 25 and 37)
Superintendent of Documents
US Government Printing Office
Washington D.C.
20402

12.4   Some further comments on MODIFICATION, MODIFICATION NOTES, and ASSOCIATED DRAWINGS

From readers' comments on the first edition of this book it is clear that a few more notes are needed on the paperwork 'mechanism' associated with 'Mods and Repairs'.

1. Prior to a modification (or a repair) the aircraft (or other equipment such as a galley) has been examined and approved by an Authority.
2. After modification the aircraft has changed, so the original approval can no longer be valid. Therefore the approval must be re-established (the CAA do this by issuing an AAN, see above).
3. The conventional way of describing what modification is to be made and of recording for the future what was done, is by engineering drawings.
4. In addition to the equipment itself, the mod will affect various documents, some of which are mandatory. These may include:

   Maintenance Manual
   Operating Manual
   Load and Balance data
   Build Standard (which lists the component parts of the individual aircraft showing the 'issue' or 'mod state' of each part at the time of its incorporation)

Master Record Index (a list of the top assembly drawings, stress reports, etc. through which a detailed knowledge of the aircraft can be traced)

5. The effect on all this hardware and paper is conveniently summarised on the Modification Note (sometimes called a Mod Sheet). This:
   (a) states what is to be done, either by a sketch on the Note itself or by reference to a drawing
   (b) records by the same method what has been done, and
   (c) lists all the documents which are needed to provide a clear understanding and record of the deviation which has taken place from the original approved article.

6. The completed Mod Note will also be the document which authorises the change to the equipment, so it must (effectively) pass through two draft stages. First it must be approved by the Design Organisation and carry their signature (see para. 12.5.5 above) and then it must receive its AAN (see para. 12.2.6) before it can be said to be complete.

7. Where the above notes refer to engineering drawings of modifications it will be clear that the drawings themselves may sometimes be changed, either to correct errors or to improve the design of the mod which they illustrate. If changing the engineering drawing involves changing the character of the modification, then everyone (including CAA and the customer) should be informed, perhaps by raised issue of the Mod Note. If the change is more minor (and judgement by the approved signatory is required here) then all that is required is a reissue of copies of the drawing to all holders of the previous issue. However, the reissue of changed drawings (or other documents) requires a systematic approach. One method is for a trusted person to physically gather and destroy the old issue copy documents and hand out new. Unless all drawings (documents) are always issued to all the departments, customers, sub-contractors, etc. this 'hand out' system can only hope to work within a very small organisation. Another, better (?) system is for the drawing office manager to write a small note saying 'By my authority drawing(s) number . . . is (are) raised from issue x to issue y. Copies of issue x must be destroyed or clearly marked as old issue'. This note should be serially numbered and its number shown against the issue letter on the drawing. This note can then be widely circulated with new drawing copies going to a more limited number of destinations. However, any workable system is satisfactory with (as always) simplicity being a prime objective.

12.5   Conclusion

Reading this book can at best set somebody on the road towards being a stressman or a structures designer. In fact, almost all the subjects he would need to study in depth are mentioned, but the qualification that no book can give is experience. Some readers who have reached this point in the book will have reached senior positions in their own field and a broad look at structural engineering will be all they want of it. Some other readers may go on to join a design organisation and acquire experience and expertise. A third group will comprise the airline engineers, on the ground or in the air, who at times want the services of a capable design organisation to make an operational improvement, design a repair or even to bring an old aircraft up to new standards. This section was written for each group to read in its own way. Group one may have read it for interest or as notes on the type of paperwork which should be emerging from the design office next door. Group three will have read it in much the same way, but with the sharper interest of knowing that if paperwork of this order is not being produced, then the improvements, repairs, etc., which they are asking for will not be given airworthiness approval For people in the middle group, this chapter like the rest of the book, is a rough definition of the terms and expressions used in their industry and perhaps an *aide-mémoire* for the first time they are out on their own, offering up the drafts of the papers needed to get approval for a change to someone else's design.

Strength of Rivets in Single Shear

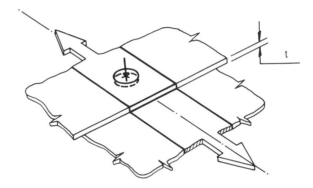

1. The figures indicate JOINT STRENGTH per RIVET in a joint which has at least two rivets.

2. The figures apply to solid, mushroom or universal head rivets in 2117 alloy, ie
British Standard SP 85
Amer Standard MS 20470 AD
and to Cherrylock Blind Rivets type CR 2249.

3. The figures are accurate enough for design purposes but not for check stressing calculations to be offered for approval.

THICKNESS t OF THE THINNER SHEET	t	RIVET DIAMETER			
		$^3/_{32}$	$^1/_8$	$^5/_{32}$	$^3/_{16}$
	.022 (24 SWG)	180			
	.028 (22 SWG)	210	295		
	.036 (20 SWG)	220	380	540	625
	.048 (18 SWG)		390	595	835
	.064 (16 SWG)		390	595	860

APPENDIX 2

Some Further Design Considerations

The following random notes suggest some minor features of detail design which might have greater significance than appears at first glance. Student engineers will gather further examples during their careers.

1. Stress concentrations and notches were discussed in para. 9.3. These notes are additions to that discussion.
The most common stress raisers are holes.

Avoid attaching equipment or secondary structure by bolting or riveting directly to areas of high stress in primary structure.

Avoid putting drain holes in areas of high stress.

Avoid putting holes for wire locking on the edge of structural members.

Avoid putting holes in the flanges of frame members (see fig. A.1). (The clipping of cable looms is the notorious cause of frame flanges being drilled without thought.)

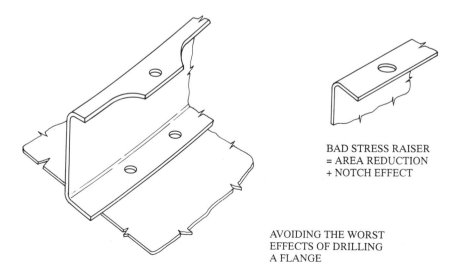

BAD STRESS RAISER
= AREA REDUCTION
+ NOTCH EFFECT

AVOIDING THE WORST
EFFECTS OF DRILLING
A FLANGE

Fig. A.1

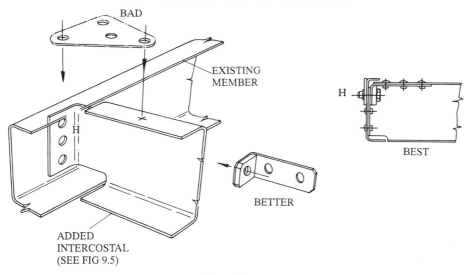

Fig. A.2

Fig. A.2 illustrates the problem of an intercostal member which has to be added to existing structure. The temptation is to join the frame flange to the intercostal flange with the triangular plate. Unless the load in the frame flange is known to be light, it is better to use a tension carrying clip bolted through hole H.

2. The next most common stress raiser is an abrupt change of section. Beware!

3. Next to the introduction of unwanted stress raisers the most common design fault in detail parts is failure to join parts together adequately. Fig. A.3 gives an example of this.

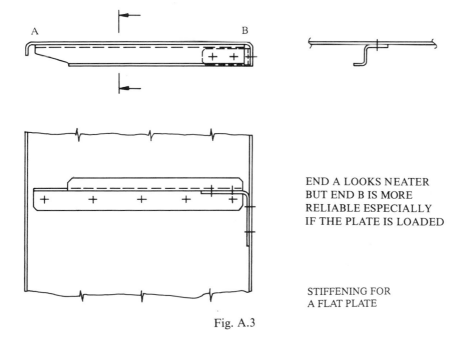

Fig. A.3

4. Beware of parts of structure which may move slightly relative to one another. (For example the floor and walls of a galley which has a top attachment to the aircraft structure as well as an attachment to the floor.) In this situation the joint between the two parts must be strong enough to resist the movement, or the parts must be properly hinged as a mechanism.

5. An error in the main text of the book was the failure to mention the *fitting factor*. This is a requirement of JAR 25.625 and means that the final fitting which joins two members together must have an additional strength reserve above its normal proof and ultimate factors (see para. 4.5). Fittings such as control surface hinge blocks, wing attachments, equipment mountings, etc., are required to have proof and ultimate reserve factors of not less than 1.15.

6. Fig. A.4 shows a hole cut in an aircraft skin (for a radio antenna perhaps) and a reinforcing or stiffening plate.

For an unpressurised aircraft the arrangement shown is satisfactory but for pressurised aircraft and in areas subject to high vibration-type loads, the arrangement is suspect on two counts.

1. Every hole in the skin is a potential start point for cracking.
2. Additional local stiffening may change the load pattern in the structure.

For a solution:

Consider the possibility of leaving the hole unreinforced – one $\frac{1}{2}$" diameter hole is probably less dangerous than the $\frac{1}{8}$" diameter holes for the rivets.

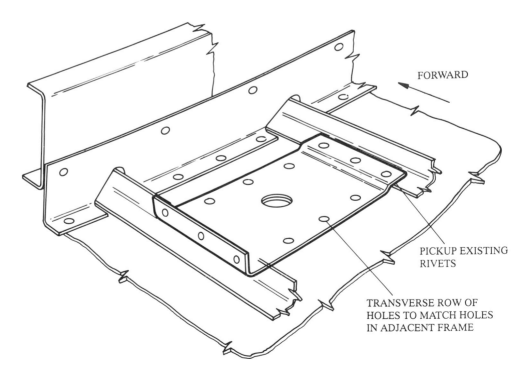

FORWARD

PICKUP EXISTING RIVETS

TRANSVERSE ROW OF HOLES TO MATCH HOLES IN ADJACENT FRAME

Fig. A.4

Consider putting a plate (same thickness as the existing skin is a 'doubler') over the whole bay i.e. picking up two stringers and two frames. This tends to avoid the first problem but still leaves the other.

Consider reinforcing the hole with a ring bonded on, possibly with Very High Bond adhesive tape.

7. After deciding on a design consult with the Approved Signatory and/or the aircraft manufacturer. He is the one who will have to correct your errors.

Index